AGRICULTURAL UNIVERSITY WAGENINGEN PAPERS
86-4 (1986)

# APPLICATION OF REMOTE SENSING TO AGRICULTURAL FIELD TRIALS

J.G.P.W. CLEVERS

Agricultural University Wageningen
The Netherlands 1986

CIP

ISBN 90-6754-089-7
ISSN 0169-345X



Printed in the Netherlands by Drukkerij Veenman b.v., Wageningen

# ABSTRACT

Clevers, J. G. P. W., 1986. Application of remote sensing to agricultural field trials (Toepassing van teledetectie bij akkerbouwveldproeven). Agricultural University Papers 86-4, 230 p., 68 eqs., 38 tables, 110 figs., 27 app., 166 refs., English and Dutch summaries.
Also: Doctoral thesis, Agricultural University Wageningen, the Netherlands.

Agricultural field trials are conducted in order to evaluate the influence of different treatments on crop characteristics, e.g. leaf area index (LAI), at different stages of growth. In the quantitative analysis large inaccuracies can occur because samples have to be small when sampling needs to be frequent and destructive.

The main aim of this study is to disclose how remote sensing can support and/or replace conventional field measurements in field trials by enabling quantitative information about an entire field trial to be obtained instantaneously, repeatedly and, above all, non-destructively.

Black and white multispectral aerial photography (with recordings being obtained in the visible and near infrared region of the electromagnetic spectrum) is shown to be the most appropriate technique for this application. The validity of canopy spectral reflectance factors, derived by a method of calibration and data correction, is demonstrated.

A simplified reflectance model for vegetation is derived for estimating LAI during the whole growing season by means of a corrected infrared reflectance factor. The latter factor is derived from the infrared reflectance by correcting for the reflectance of the background (soil and yellow leaves). Statistical analysis of several field trials reveals that the presence of treatment effects can be shown with larger power and that coefficients of variation are smaller for the LAI estimated by means of the simplified reflectance model than for the LAI measured in the field.



Free descriptors: field trials, remote sensing, multispectral aerial photography, reflectance factors, simplified reflectance model, estimating crop characteristics, statistical analysis.

CONTENTS

# LIST OF SYMBOLS AND ABBREVIATIONS

| | |
|---|---|
| B | : Soil cover |
| CV | : Coefficient of variation |
| F0,F1 | : Fungicide treatments |
| L | : Leaf area index |
| LAI | : Leaf area index |
| MSP | : Multispectral photography |
| MSS | : Multispectral scanning |
| N1..N5 | : Nitrogen levels |
| $r$ | : Total measured reflectance |
| $r'$ | : Corrected reflectance |
| $r_g$ | : Total measured reflectance in a green passband |
| $r_{ir}$ | : Total measured reflectance in an infrared passband |
| $r'_{ir}$ | : Corrected reflectance in an infrared passband |
| $r_r$ | : Total measured reflectance in a red passband |
| $r_s$ | : Reflectance of the soil |
| $r_{s,g}$ | : Reflectance of the soil in a green passband |
| $r_{s,ir}$ | : Reflectance of the soil in an infrared passband |
| $r_{s,r}$ | : Reflectance of the soil in a red passband |
| $r_v$ | : Reflectance of the vegetation |
| $r_{v,g}$ | : Reflectance of the vegetation in a green passband |
| $r_{v,ir}$ | : Reflectance of the vegetation in an infrared passband |
| $r_{v,r}$ | : Reflectance of the vegetation in a red passband |
| $r_{\infty,ir}$ | : Parameter describing the asymptotic value for the corrected infrared reflectance |
| R1..R3 | : Rotations |
| RSS | : Residual sum of squares |
| $s^2$ | : Residual variance |
| S1..S3 | : Plant densities |
| $\bar{x}$ | : Mean value of x |
| $\hat{x}$ | : Estimated value of x |
| Z1,Z2 | : Sowing dates |
| $\alpha$ | : Parameter related to a combination of extinction and scattering coefficients |
| $\mu$ | : mean |
| $\lambda$ | : wavelength |
| $\sigma$ | : standard deviation |
| $\sigma^2$ | : variance |
| $\varepsilon$ | : expectation |

# 1 INTRODUCTION

## 1.1 Aim of this study

For many years much experience has been acquired in collecting, analysing and interpreting data from field trials. During the growing season of agricultural crops, many plant characteristics may be ascertained in field trials. Often an investigator has the dilemma of deciding which relevant characteristic can be ascertained quickly and accurately. Both conditions are often contradictory. The accurate measurement of plant characteristics is very time-consuming and may also require intensive use of labour and apparatus.

One aim of this study is to identify a method that can reduce inaccuracies in field trial analysis. Attention was focussed on remote sensing techniques. Remote sensing techniques enable information about a whole field trial to be obtained quantitatively, instantaneously and, above all, non-destructively. A second aim is to identify how remote sensing can support and/or replace conventional field measurements in field trials. The data must be analysed according to the design of the field trial and here the main aim is to reduce inaccuracies.

## 1.2 Background to the present study

Over many years much experience has been acquired in performing and analysing agricultural field trials with arable crops. These trials are indispensable for testing aspects such as new varieties, new fertilizers or pesticides, and new cultivation techniques in general. In all cases the pros and cons of new practices have to be investigated before there is an incentive for introducing them into practical farm management.

The agronomist is not only interested in the yield, but also in vegetation characteristics during the growing season. These latter observations can provide information about factors that could explain differences in yield and they may also indicate in what way farm management can be further improved. An example is nitrogen nutrition in winter wheat in the Netherlands. Field trials showed that an increased nitrogen nutrition boosts the amount of green biomass, which means an increased potential yield. Above a certain level of nitrogen, lodging prevented the yield from reaching its optimum. If only this lodging could be suppressed, yield could still be increased by increasing nitrogen levels. In order to do this, new short-stemmed cultivars have been introduced and growth retardants have been applied to control the length of the stems. As a result, the yield of winter wheat in the Netherlands has increased considerably during the past decades.

To obtain data on vegetation characteristics during the growing season the investigator has two approaches at his disposal. The first one yields qualitative information. Walking through the fields he estimates soil cover, average plant height or the amount of biomass; he may judge the plots on the basis of features such as leaf colour or leaf orientation (disease), and assign a subjective grade to each plot. The advantages of this method are:

1. It is a rapid method. The investigator has the results at his disposal immediately.
2. It offers information about an entire plot (= smallest experimental unit) of a field trial.
3. As a result of 2, the same plants are involved in the judgement each time.

This method has some severe disadvantages:

1. Observations are often arbitrary and highly subjective.
2. Judging a number of plots takes time, during which the features of the plants may change (for instance because of wind or drought).
3. The judgement of the observer will not be consistent, but will change during the day in a random, non-systematic way, or because of other factors, e.g. tiredness.

The second approach yields quantitative information. This method is based on taking samples from each plot. In this way important plant characteristics such as leaf area index (LAI; defined as the total one-sided green leaf area per unit soil area), fresh weight, dry matter weight and dry matter content may be estimated. The main advantages of this method are:

1. Data are quantitative.
2. Data are gathered objectively.

Consequently, data from various plots are more comparable and results are more comparable with results obtained by other investigators. Serious disadvantages are:

1. This form of data gathering is destructive. Typically, sampling may take place fortnightly and as a result, only a relatively small proportion of the plants available can be harvested. Alternatively, the total area of a plot would have to be very large; the entire field trial would become uneconomically large. When sample size is small the measurements have a large variability (variance) due to the influence of e.g. soil variability and individual plant variability (e.g. Spiertz & Ellen, 1978; Daughtry & Hollinger, 1984).
2. Samples, especially large ones, require much labour for analysis. Therefore it is normally several days before results are available.
3. Each time, different plants from the same plot have to be harvested. This increases the variability between sampling dates.

The usual method of acquiring quantitative data in field trials is a multiple sampling method. The first measurement involves ascertaining total fresh weight of the sample. Measurement of this characteristic often involves the largest inaccuracy (e.g. because of local differences in growth conditions). A secondary measurement is dry matter content or nutrient content, which can be ascertained by analysing a subsample. This may also be done for individual plant parts. Another secondary measurement is the leaf area per unit leaf weight (i.e. the specific leaf area). To calculate the LAI, the fresh weight of the leaves of the whole sample is used, together with the specific leaf area. Within one plot, variations in dry matter and nutrient content or specific leaf area are normally much smaller than variations in total fresh weight.

The determination of LAI involves some specific problems. First of all, after harvesting, the plants have to be separated in the laboratory into leaf blades and other plant parts and the leaf area has to be measured. This whole procedure is time-consuming, and meanwhile the leaves may shrink. This may severely reduce the accuracy of the measurements. During senescence, leaves have to be separated into green and yellow leaves, because LAI only refers to green leaves. This distinction between green and yellow is very subjective and will vary considerably from person to person because often a leaf is partly green and partly yellow and the transition is gradual. This makes the measurement of LAI subjective.

Given the drawbacks of the multiple sampling method and the subjectivity and inaccuracy of ascertaining specific leaf area, the overall variation of LAI measurement will be larger than the variation in fresh or dry matter weight. For instance, the coefficient of variation (CV, for definition see section 2.5.1) in field trial 116 in 1982 (chapter 7) for dry matter weight ranged from 0.11–0.23, the CV for the specific leaf area ranged from 0.05–0.15 (not listed) and the CV for LAI ranged from 0.21–1.30. Still, LAI is regarded as being a very important plant characteristic because photosynthesis takes place in the green plant parts and LAI is relatively simple to measure. Ideally, the agronomist would rather measure the total photosynthetic activity.

In order to reduce this sampling inaccuracy in agricultural field trials one would like to have a method at one's disposal which:

1. offers quantitative information, enabling results to be objectively analysed and compared
2. offers instantaneous information for the entire area
3. is non-destructive, thus enabling frequent measurements on the same plants
4. is applicable to a relatively large area, yielding a mean value for an entire plot (e.g. of at least 50 $m^2$), thus minimizing the influence of soil variability and individual plant variability.

In order to identify such a method, which was one aim of this study, attention was focussed on techniques of remote sensing, because this field seems to offer great potential for application to agronomy.

During the past decades knowledge about remote sensing techniques and their application to fields such as agriculture has improved considerably. The work done by Bunnik (1978) formed the starting point for this study. He demonstrated the possibilities of applying remote sensing in agriculture, particularly with regard to its relation with crop characteristics such as soil cover, LAI and dry matter weight. Although much research has been done on the relationship between crop characteristics and remote sensing measurements, the application of remote sensing to field trials itself has never been investigated extensively. Nevertheless, field trials are sometimes used in remote sensing research work, for they provide a wide range of values on the crop characteristic to be investigated, because different treatments have been applied. For ascertaining the relationships between remote sensing measurements and crop characteristics accurately, such a wide range is preferable (see, e.g. Hatfield, 1983).

## 1.3 Main aspects studied

The hypothesis that inspired this study is that remote sensing can be used in agricultural field trials. The major prerequisites for the technique used for this particular application are:

a. It should offer characteristic spectral information about the vegetation and small differences within one crop in field trials should be detectable.
b. It should not be too expensive to use, because it has to be applied repeatedly during the growing season (temporal information).
c. The spatial resolution (resulting from a combination of a recording and a processing system) should be in the order of a few square metres or better, given that plots in agricultural field trials are usually small.
d. It should permit a large area (up to several hectares) to be recorded in a short time period (a matter of minutes).
e. It should be possible to have the results available within one or two days.

The first stage in finding the ideal remote sensing technique for use in agricultural field trials is to ascertain:

1. the wavelength region and, in particular, the spectral bands that should be applied for obtaining characteristic information. (See prerequisite a, above.)
2. the sensor and platform that are most appropriate for monitoring agricultural field trials. (See prerequisites b, c and d.)
3. the processing system that is not too time-consuming and labour intensive. (See prerequisite e.). Furthermore,
4. the sensor and processing system have to be calibrated so that the information obtained is influenced only by the object being studied.

After ascertaining the technical specifications above, the next stage is to ascertain the significance of remote sensing measurements for crop information: points 5-7 below. Each of these questions opened up a line of research that was followed in this study.

5. What is the agronomic meaning of the variables measured by remote sensing techniques?
6. How can remote sensing measurements be used for estimating conventional variables measured in the field?
7. Can remote sensing measurements totally replace field measurements?

## 1.4 THE RESEARCH STRATEGY

The research was carried out at the ir. A.P. Minderhoudhoeve, experimental farm of the Wageningen Agricultural University (the Netherlands), situated in one of the new polders, Oost-Flevoland (figure 1.1), which was reclaimed about 30 years ago. The new polders are flat, uniform and highly productive agricultural lands with a loamy topsoil. Measurements were gathered during the 1981-1983 growing seasons (cf. chapter 6).

First of all, in line with points 1 and 2 mentioned in section 1.3, the spectral bands, sensor and platform, appropriate for application at agricultural field trials were selected, primarily by evaluating results from literature. Applicable systems then were tested at the ir. A.P. Minderhoudhoeve and the most appropriate one for use in agricultural field trials was selected. Also the processing system was tested in order to obtain quantitative information about agricultural crops (see point 3 of section 1.3). Next, ways of calibrating the whole system (sensor plus processing) to produce calibrated variables that can be analysed multitemporally were investigated (see point 4 of section 1.3).

Secondly, the agronomic meaning of the variables measured using the selected sensor system was ascertained (see point 5 of section 1.3) and it was investigated whether these variables can be determined with relatively smaller variance than the conventional variables. Aspects investigated included the ability to ascertain treatment effects with larger 'power', i.e. improving the probability that the null hypothesis (that the treatment has no effects) will be rejected on the ground of remotely sensed data. Both data-gathering methods (remote and in the field) contain inaccuracies, and in this research an attempt was made to assess each method as objectively as possible, by using statistical techniques.

Since the agronomist is currently still interested in conventional field data, the variables measured by remote sensing were related to the conventional field data. In this study the possibilities of estimating the latter by the former were investigated (see points 6 and 7 of section 1.3). The possibilities of applying some index or model from literature for estimating crop characteristics were evaluated. Since none were found to be suitable for the purpose of this study, an appropriate model was derived by adopting a few legitimate assumptions, the parameters being estimated empirically. For this monograph and its specific application to field trials, the ideal model is one that is simple and requires the least number of input variables. The practical applicability of the model is very important. For instance, if the agronomist has to ascertain a leaf angle distribution (necessary for some existing models), he will probably prefer to collect the conventional field data as he has always done.

Finally, the above model and hypotheses were tested and investigated with actual data obtained in the field and by remote sensing at the ir. A.P. Minderhoudhoeve. Most data were analysed according to the design of the field trial, by applying an analysis of variance. Measures of inaccuracy obtained in this way were used for comparison: remote sensing measurements were compared with conventional crop characteristics, and also crop characteristics estimated

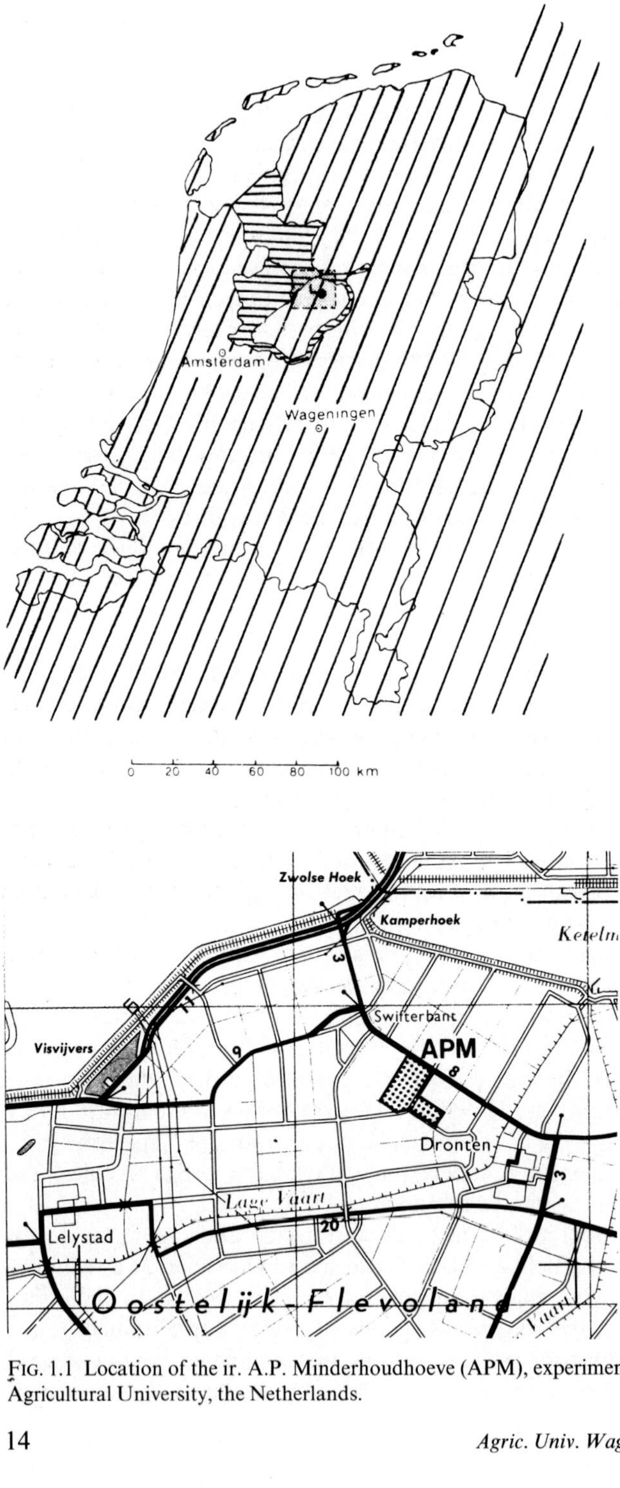


///: land
═: enclosed water

FIG. 1.1 Location of the ir. A.P. Minderhoudhoeve (APM), experimental farm of the Wageningen Agricultural University, the Netherlands.

from remote sensing measurements were compared with crop characteristics measured in the field.

## 1.5 Organization of the thesis

In chapter 2 the available sensor systems are evaluated; this evaluation reveals that multispectral aerial photography (MSP) is a very appropriate method for use in agricultural field trials. The optimal spectral passbands were selected from those available with the Daedalus multispectral scanner. Multispectral aerial photography and its processing and calibration are comprehensively described in chapter 3, which concludes with a comparison between reflectance factors obtained by multispectral photography and reflectance factors measured with other sensors (e.g. radiometers in the field).

Chapter 4 gives definitions for reflectance, reflectance factor and reflected radiance, which are important for the sensor system used. The main crop characteristics that may be measured with the recommended system (MSP) are described and the influence of the soil is considered. Finally, several indices and reflectance models from literature for estimating crop characteristics and their applicability in this research are considered.

Chapter 5 presents new models for estimating soil cover and LAI. These models are compared with an existing, more complicated model (the SAIL model).

Chapter 6 describes the conventional methodology applied in field trials and the design of several field trials that have been investigated, plus the statistical interpretation of the field data and the procedures used to compare and relate spectral measurements to field data.

Chapter 7 presents the main results of comparing reflectance measurements with crop characteristics and of estimating the latter by the former for several field trials. A more extensive enumeration of results is given in the appendices (presented as tables and figures).

Finally, chapter 8 gives final remarks as well as recommendations for future work; chapter 9 gives the main conclusions of this research.

# 2 EVALUATION OF REMOTE SENSING SYSTEMS

## 2.1 Introduction

After giving some general information about remote sensing, relevant literature will be reviewed to reveal the remote sensing systems that fulfil the requirements given in section 1.3 for suitability for use in field trials. These systems were tested at the ir. A.P. Minderhoudhoeve, so that the most appropriate one could be selected. Finally, results from a multispectral scanning system, also tested at the ir. A.P. Minderhoudhoeve and used to obtain an optimal choice of channels (passbands), are given.

## 2.2 General aspects of remote sensing

In this study the applicability of remote sensing techniques to agricultural field trials was investigated. Since many agronomists for whom the results of this study could primarily be important may not be familiar with remote sensing, this chapter begins with very general information about this technique.

Remote sensing enables one to acquire information about an object from a distance, that is without being in contact with the object. Sensors in airborne or spaceborne platforms operate in the electromagnetic spectrum (figure 2.1).

Electromagnetic radiation from the sun that reaches the earth's surface will hit an object (figure 2.2). This may result in one of three interactions:

- transmission of radiation by the object because the object is wholly or partly transparent to this radiation;
- absorption of radiation by the object, i.e. radiation is retained by the object and may be used for certain internal processes (e.g. photosynthesis);
- reflection of radiation at or near the surface of the object.

In addition to the radiation reflected, the radiation emitted by an object on

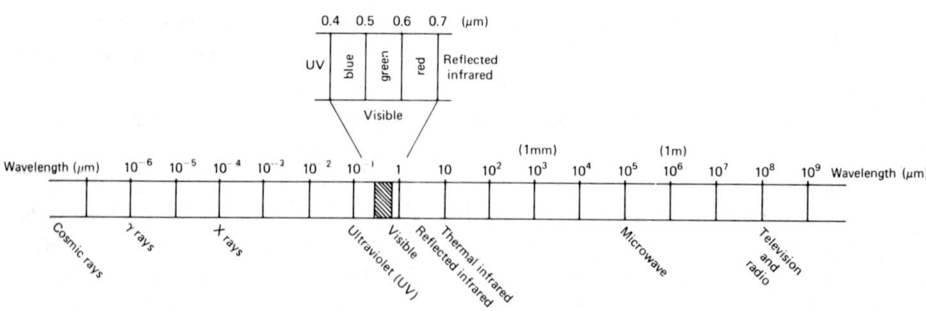


FIG. 2.1 The electromagnetic spectrum (from: Lillesand & Kiefer, 1979).

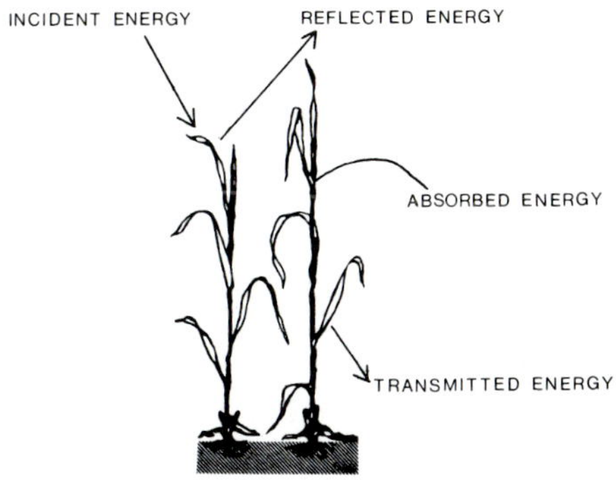


FIG. 2.2 Basic interactions between electromagnetic energy and vegetation.

the earth's surface may be remotely sensed. In the latter case we speak of thermography, since the emission is thermal infrared or heat radiation.

Another method, frequently applied, involves an active system operating in the microwave region of the electromagnetic spectrum. This system has its own source of energy radiation and will register the reflected energy (e.g. microwaves; figure 2.1).

The energy interaction is often specific for a certain object. This specificity may be used for distinguishing objects or for ascertaining certain characteristics of an object. The energy interaction may be equal for different objects in one part of the electromagnetic spectrum, but different in another part of the spectrum. Knowledge about the reflectance pattern is referred to as spectral information. In this respect it is important to know the radiation wavelengths to which a single sensor is sensitive (its spectral sensitivity). Spectral resolution is a measure of both this spectral sensitivity and the discreteness of the bandwidths of the spectral wavelength ranges (after: Swain & Davis, 1978), indicated as spectral bands, passbands or channels.

Information concerning any change in the spectral characteristics of an object over time may also be important in many applications. This is referred to as temporal information. In this respect it is important to know whether a sensor can be applied under all weather conditions, and how expensive reiterative observations are.

The third kind of information is spatial information, which is concerned with characteristics dependent on location. It is important to know the limitations of a sensor system with respect to spatial detail (its spatial resolution).

As well as the spectral, temporal and spatial resolution of a sensor, the radiometric resolution must be known to answer questions 1 to 4 posed in section 1.3. This radiometric resolution is defined as the smallest discriminable difference of signal output of a sensor.

All radiation detected by remote sensors has travelled through the atmos-

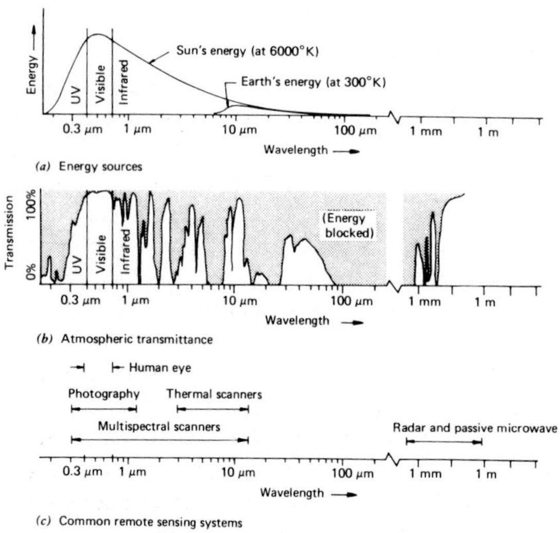


FIG. 2.3 Spectral characteristics of energy sources, atmospheric effects, and sensing systems (from: Lillesand & Kiefer, 1979).

phere. The atmosphere may modify or contribute to the radiation coming from the earth's surface. Different kinds of particles present in the atmosphere will absorb and scatter the radiation passing through it, and this has to be taken into account. It is especially important to remember that the atmosphere has limited transparency in certain parts of the electromagnetic spectrum (bands) because there is a strong absorption of energy by the atmosphere in those bands (figure 2.3). This will restrict the application of remote sensing to certain 'windows' in the electromagnetic spectrum.

As well as airborne and spaceborne sensors, earth-bound sensors may be used. These are especially useful for detailed studies where a limited number of measurements is required, or for supplying reference data for checking other remotely sensed data. These sensors are especially important for checking the calibration of other remote sensing systems and/or for atmospheric correction.

## 2.3 REVIEW OF RELATED RESEARCH ON SEVERAL SENSORS

Below, a selection from the extensive literature on remote sensing research is reviewed, to show the potential of the various apparatus available.

Remote sensing measurements can be carried out at different levels (altitudes), ranging from measurements in the field up to measurements from space.

In order to determine the spectral reflectance of individual leaves or plants laboratory measurements are carried out (e.g. Gausman, 1982; Gausman et al., 1973; Horler et al., 1983; Tucker, 1980). Such measurements are very labour intensive, time-consuming and sensitive to the variability of individual leaves and plants. So, they are not suitable for monitoring field trials.

For determining the relationship between remote sensing measurements and crop characteristics, hand-held radiometers are often used (e.g. Aase et al., 1984; Ahlrichs & Bauer, 1983; Hatfield et al., 1984; Holben et al., 1980; Idso et al., 1981; Markham et al., 1981; Miller et al., 1984; Milton, 1980; Pearson & Miller, 1973b; Pearson et al., 1976; Pinter et al., 1981; Steven et al., 1983; Tucker, 1980; Tucker et al., 1973, 1979, 1980). These hand-held radiometers can only acquire reflectance measurements in a few passbands at discrete locations. For acquiring spectral reflectance measurements on site in narrow passbands over a wide range of wavelengths, spectroradiometers are placed on elevated platforms (e.g. Brown & Ahern, 1980; Janse & Bunnik, 1974; Pearson & Miller, 1971; Tucker et al., 1973; Verhoef & Bunnik, 1974). Although they are unsuitable for use on large areas within a short time period, hand-held radiometers and spectroradiometers may be of some use in agricultural field trials. Therefore they were tested at the ir. A.P. Minderhoudhoeve (section 2.4). Results of the measurements are used in section 3.13.

Another level for acquiring remote sensing measurements is from an aerial platform. In this way larger areas can be recorded. The various airborne sensors operate in different parts of the electromagnetic spectrum (figure 2.3). The spatial resolution of the microwave sensors currently available is too low for use in field trials. Moreover, these sensors are too expensive for frequent recordings to be made during the growing season. Thermal sensors are often used in combination with multispectral scanning systems. Thermal sensors record the emitted (thermal) infrared radiation from a surface (offering, for instance, information about evaporation of crops which may indicate differences in stress, such as drought). Since these sensors offer no direct information about soil cover, LAI or dry matter weight (which are the main crop characteristics of interest), their use in agricultural field trials, and therefore their relevance to this study, was ruled out. Multispectral scanners record digitally the reflected visible and infrared radiation in several wavelength bands (passbands). They have a high spectral and a moderate (a few square metres) spatial resolution. Some research has been done by using aerial multispectral scanners (e.g. Aase et al., 1984; Bunnik et al., 1977; Hatfield et al., 1982; Wardley & Curran, 1984). Since, in principle, a multispectral scanner is applicable to agricultural field trials, this device was also tested at the ir. A.P. Minderhoudhoeve (next section). Systems for aerial photography also enable recordings of the reflected visible and infrared radiation to be made. The spatial resolution of photographic systems is high, being mainly confined by image motion and by the aperture of the device (the densitometer) used for measuring densities in photographs; cf. section 3.5. To date, the calibration of aerial photography has proved problematic, in particular because of the analog data registration (e.g. see Curran, 1980, 1981, 1982a, 1982b, 1983; Graham, 1980; Kannegieter, 1980; Ross, 1973; Sievers, 1976). If this calibration problem could be overcome, aerial photography would be particularly useful in agricultural field trials. Therefore this method of remote sensing was also tested at the ir. A.P. Minderhoudhoeve and described in the next section.

In order to map or classify large areas, spaceborne sensors (satellites) can be used (see Gray & McCrary, 1981a, 1981b; Heilman & Moore, 1982; Markham et al., 1981; Pollock & Kanemasu, 1979; Rouse et al., 1973; Tucker et al., 1985). Their spatial resolution (> 10 by 10 metres) is too low for use in field trials, therefore they will not be considered further.

One of the main results of the work done by Bunnik (1978) was the identification of five wavelengths based on optimum information about variation in relevant crop characteristics. These wavelengths were: one in the green at 550 nm, one in the red at 670 nm, one in the near infrared at 870 nm and two in the water absorption region – one at 1 650 nm and the other at 2 200 nm. Recordings in the water absorption region taken from any aerial platform are difficult to apply, because of the modification by water vapour in the atmosphere. With aerial photography, recordings in the water absorption region cannot be made, since no film material is sensitive to that radiation. Bunnik also discussed the bandwidths acceptable for registering reflectance of crops. This bandwidth was determined by a maximum variance in the reflectance of vegetation with variable crop and soil properties. In the visible region (550 nm and 670 nm) the bandwidth should be small (about 20 nm), but it has to be a compromise between the required width and the low signal level caused by the generally low reflectance of green vegetation, especially in the red. In the infrared region at 870 nm the band can be wider (e.g. 100 nm), provided that the water absorption at 940 nm is excluded.

In the literature there is a certain consensus that bands in the green, red and near infrared regions are optimal if information about vegetation is to be obtained (e.g. Kondratyev & Pokrovsky, 1979).

## 2.4 Selection of sensor and platform

In this section the efficacy of four sensor systems that were available and could be used in field trials is evaluated. Each of these systems was tested for their usefulness in agricultural field trials at the ir. A.P. Minderhoudhoeve (figure 1.1) during the 1981-1983 growing seasons. These systems were a hand-held radiometer, a field spectroradiometer, aerial multispectral scanning and multispectral aerial photography. The hand-held radiometer is a portable instrument for carrying out reflectance measurements at discrete locations in a few passbands. The field spectroradiometer is an instrument that records in many narrow spectral bands, yielding the spectral distribution of radiant energy. Both instruments are used for verifying measurements from airborne sensors. The pros and cons of aerial multispectral scanning and aerial multispectral photography were evaluated.

The hand-held radiometer, which carries out reflectance measurements in

TABLE 2.1 Specifications of the filters used with the hand-held radiometer.

| central wavelength (nm) | maximum transmittance | wavelength (50% rel. transmittance) | bandwidth |
|---|---|---|---|
| 577 | 45% | 565–590 nm | 25 nm |
| 660 | 70% | 645–675 nm | 30 nm |
| 840 | 45% | 834–846 nm | 12 nm |

three passbands, was constructed by the Technical and Physical Engineering Research Service (TFDL), Wageningen. Uenk (1982) has described a similar instrument. The receiver of the radiometer consists of two photo-electric cells mounted in the centre of a rotating drum, one measuring the incoming radiance (the total of sunlight and skylight) and the other measuring the reflected radiance. The surface of the drum contains three optical filters, whose specifications are given in table 2.1. The filters were chosen on the basis of the results obtained by Bunnik (1978) and on the channels of the Daedalus multispectral scanner (channels 5, 7 and 9). The spectral passbands are illustrated in figure 2.4. The device measures incoming radiance through a so-called cosine-corrected sphere. Several apertures can be used to measure the reflected radiance. Data are stored (and processed to some degree) by a pocket calculator, which is interfaced with the radiometer.

We tested the usefulness of the hand-held radiometer for agricultural field trials at the ir. A.P. Minderhoudhoeve. The radiometer was calibrated by taking reflectance measurements using the reference targets (artificial targets with known reflectance factors) described in section 3.12. The instrument was held about one metre above the object (figure 2.5), whilst avoiding shadow on the object. The measured area of the object was about half a square metre. In order to obtain an average plot value, 6 measurements were carried out per plot. Measuring one plot (of a field trial) in this way took about 3 minutes.

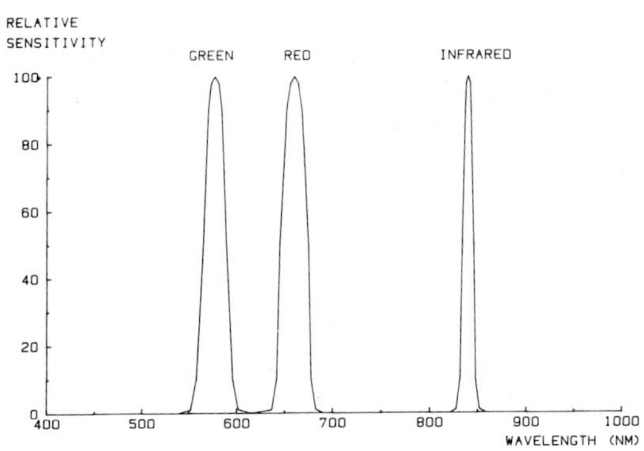


FIG. 2.4 Spectral passbands of the hand-held radiometer.

FIG. 2.5 The hand-held radiometer.

The field spectroradiometer tested has been described by Bunnik (1978). It was designed and constructed by the TNO-TH (TPD) Institute of Applied Physics, Delft. To keep the measurement conditions as constant as possible, the instrument simultaneously measures an object and a diffuse reflecting reference target, both exposed to the same irradiation. The reference target comprised a panel sprayed with Eastman Kodak White Reflectance Paint (based on $BaSO_4$). This panel was calibrated in the laboratory with a standard panel with known absolute reflectance. The spectroradiometer is constructed in such a way (cf. Bunnik, 1978) that the object field of view equals about $8° \times 20°$ and the reference field of view is 5 mrad. Two mirrors, oriented at an inclination of 45°, were used to observe the object and the reference panels from perpendicularly above (figure 2.6). The distance between entrance pupil and object varied from 5 m to 6 m, depending on object height. At object level this resulted in an area varying from 0.70 m by 1.75 m to 0.85 m by 2.10 m. Both the object or reference panel had to be free of shadow from the spectroradiometer. By using this instrument, spectral measurements can be obtained in 153 passbands within a wavelength interval from 361 nm at the end of the ultraviolet region to 2 360 nm in the reflective infrared region. A scan through the whole spectrum is obtained in about 2 minutes. The output signals of the spectroradiometer are stored on magnetic tape after digitalization by a datalogging system. The data obtained are then processed by computer, which makes a correction for the absolute reflectance of the reference panel and compensates for systematic differences between the output of the object optics and of the reference optics. Some specifications of the field spectroradiometer are given in table 2.2 (after

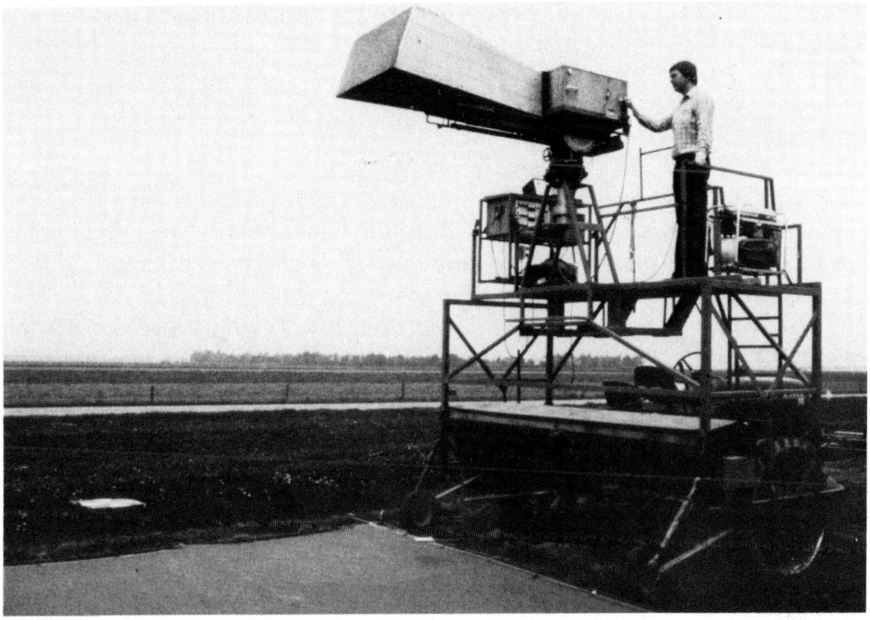

FIG. 2.6 The field spectroradiometer.

Bunnik, 1978). To test the usefulness of the spectroradiometer for agricultural field trials at the ir. A.P. Minderhoudhoeve it was mounted on a tractor for mobility (figure 2.6). It took about 5 minutes to take one measurement in a certain plot and then move to the adjacent plot.

Multispectral scanning (MSS) recordings of field trials at the ir. A.P. Minderhoudhoeve were obtained by means of a 10-channel 'Daedalus Multispectral Scanner', model 1240/1260. The missions were performed by Eurosense B.V. and financed by the BCRS (= Begeleidings-Commissie Remote Sensing). The digital data were recorded on high density tape in the aeroplane (a Dornier Skyservant DO.28 D-1) and afterwards corrected and resampled for panoramic distortion and then converted into computer-compatible tape (CCT). The CCTs were processed in the computer colourgraphic system of the Wageningen Agri-

TABLE 2.2 Specifications of the spectral resolution of the field spectroradiometer (after Bunnik, 1978).

| interval | detector | spectral range | bandwidth |
|---|---|---|---|
| 1 | Si | 361– 753 nm | 17 nm |
| 2 | Si | 629–1226 nm | 25 nm |
| 3 | PbS | 1165–2360 nm | 42 nm |

cultural University. Recordings were obtained in the following channels:
channel 4 : 500–550 nm (green) [in 1982 only]
channel 5 : 550–600 nm (green)
channel 6 : 600–650 nm (red)
channel 7 : 650–700 nm (red)
channel 8 : 700–790 nm (reflective infrared)
channel 9 : 800–890 nm (reflective infrared).

No recordings were obtained at wavelength bands below 500 nm because of the large influence of the atmosphere in this region of the electromagnetic spectrum.

The radiance reaching the sensor is converted into a digital value by electronic energy detectors, and may be recalculated to radiance in a direct manner by means of an internal calibration system. Each digital value (or radiance value) is an integrated value over a small surface area in the scene to be analysed (for a complete description of such a system, see remote sensing handbooks, e.g. Manual of Remote Sensing, 1983). The digital dataset corresponds to a 2-dimensional grid. A rotating mirror senses the ground surface along a series of parallel scan lines running perpendicular to the flight direction. By the forward motion of the aeroplane new strips of the surface to be covered by successive scan lines will be measured. Each cell of this grid is called a picture element (pixel). This type of scanning system is supposed to be a well calibrated system. The minimum size of a pixel at ground level determines whether multispectral scanning can be used for agricultural field trials.

In processing the MSS data it is important to determine accurately a mean value for radiance reflected by each plot in a field trial. The smallness of the net plots can be a problem. In our tests they were only 3 metres wide and 15 to 20 metres in length. The minimum height for recording with the available scanning system is about 1 500 feet (about 460 metres) for proper navigation, to minimize the movements of the plane and to reach the correct ratio of scanning speed to flight speed. The scanning system has an instantaneous field of view of 2.5 mrad (0.143°) (circular) and a total field of view of 86° (figure 2.7). The instantaneous field of view is defined as the smallest plane angle over which an instrument (e.g. a scanner) is sensitive to radiation; the total field of view is defined as the overall plane angle in the across-track direction (y-direction) (after: Swain & Davis, 1978). At an altitude of 1 500 feet this implies that the pixel is circular with a diameter of 1.14 metres at a view angle of 0° (nadir viewing) and that the pixel becomes elliptical with increasing view angle up to a pixel with axes of 1.56 metres and 2.14 metres in the principal directions of the ellipse at maximum view angle. The shorter axis of the ellipse (x-direction) was arranged to be nominally parallel to the shorter side of the plots (= 3 metres). In our tests, only one scan line covered the plots without interference from the edges of the plots. Moreover, the forward movement of the plane and its other movements (roll, yaw, pitch) mean that the scan lines are never perfectly perpendicular to the flight direction (figure 2.8). Thus there were only a few pixels on a net plot and consequently their location had to be ascertained accurately.

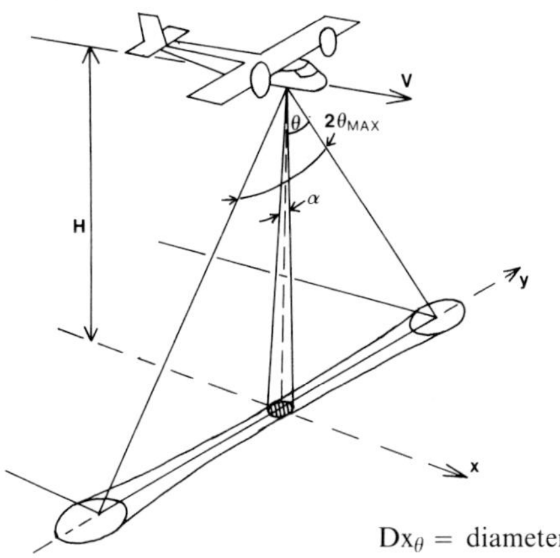


| | |
|---|---|
| $\theta$ | = view angle |
| $\alpha$ | = instantaneous field of view |
| $2\theta_{max}$ | = total field of view |
| V | = flight speed |
| H | = flight height |
| y | = y-direction (= scan direction) |
| x | = x-direction (= flight direction) |
| $Dx_\theta$ | $= \dfrac{H \cdot \alpha}{\cos\theta}$ |
| $Dy_\theta$ | $= \dfrac{H \cdot \alpha}{\cos^2\theta}$ |
| $S_\theta$ | $= \pi \cdot \dfrac{Dx_\theta}{2} \cdot \dfrac{Dy_\theta}{2}$ (= elliptical) |

$Dx_\theta$ = diameter of pixel in x-direction, depending on view angle $\theta$.
$Dy_\theta$ = diameter of pixel in y-direction, depending on view angle $\theta$.
$S_\theta$ = pixel area a sground resolution, depending on view angle $\theta$.

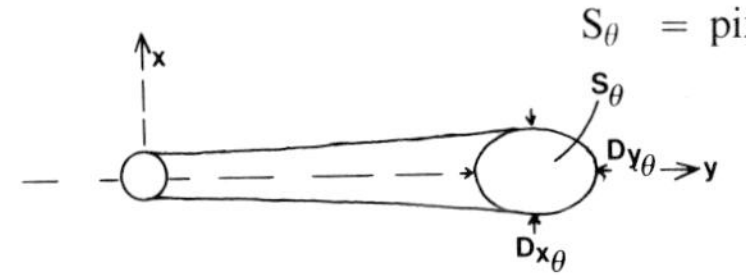


FIG. 2.7 Schematic presentation of scanning geometry of a multispectral scanning system.

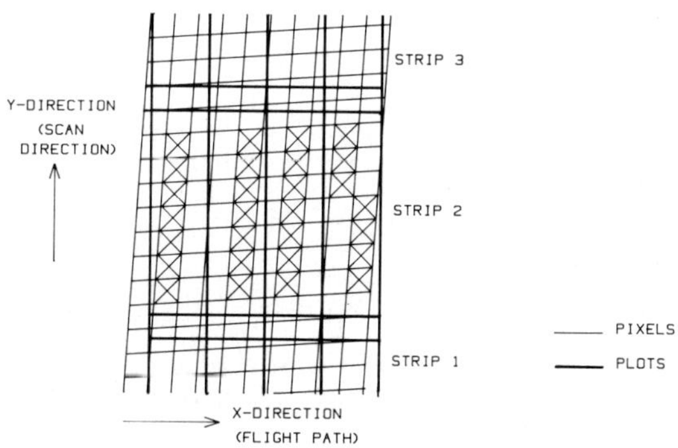


FIG. 2.8 Schematic presentation of possible positions of pixels across the plots of a field trial. Pure pixels are indicated by a cross.

Moreover, their location cannot be influenced afterwards (cf. aerial photography).

A mission using such a system is too expensive (in the order of f 40 000 per mission for the CCTs) to be used very frequently during the season. This rules out its use for repetitive overflights during the whole growing season. If only one or two overflights are required these systems would be feasible. In the following section scanner data will be used to choose the optimal spectral passband, using results obtained from field trials at the ir. A.P. Minderhoudhoeve.

Systems of aerial photography enable recordings of the reflected visible and infrared radiation to be acquired. The spatial resolution will primarily depend on lens type, altitude, film material and image motion. By using an appropriate micro-densitometer or densitometer with small aperture, quantitative measurements with high spatial resolution may be obtained (see section 3.5). Moreover, the measurements can be carried out at exactly determinable locations of an object. The camera equipment can be installed in small planes if small non-metric cameras are used (e.g. 70-mm aerial photography).

The potential of integral tripack colour infrared film (Aerochrome 2443) with its merit of simultaneous exposure and registration of three spectral bands, i.e. green, red and infrared, was studied. Its use was discounted, as problems were foreseen in extracting spectral reflectance data from a sensor of modest sensitivity (E.A.F.S. = 40 without the use of filters, where E.A.F.S. = Effective Aerial Film Speed as determined by a threshold exposure response measured at a density of fog + 0.3; fog being the density of an unexposed emulsion after processing, added to the density of the plastic base of the film). In addition, inter-image effects may occur (Egan, 1985). Moreover, spectral resolution is low (figure 2.9). The infrared band (cyan dye layer) is very wide, and sensitive even for green and red radiation. Therefore, separation into green, red and infrared is very complicated (see e.g. Scarpace, 1978; Scarpace & Friederichs, 1978; Scarpace & Quirk, 1982).

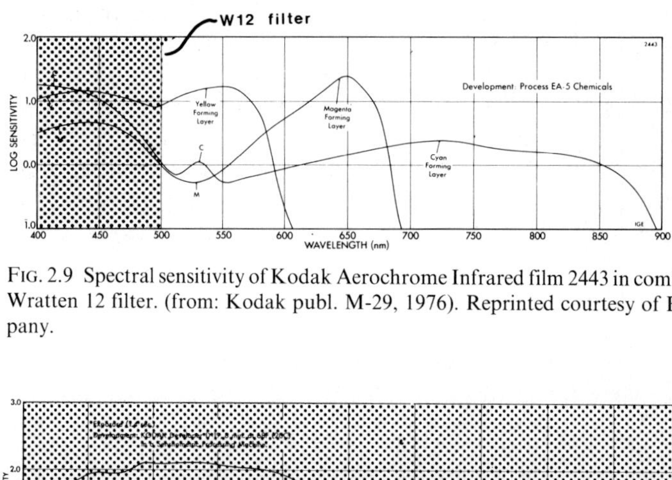


FIG. 2.9 Spectral sensitivity of Kodak Aerochrome Infrared film 2443 in combination with a Kodak Wratten 12 filter. (from: Kodak publ. M-29, 1976). Reprinted courtesy of Eastman Kodak Company.

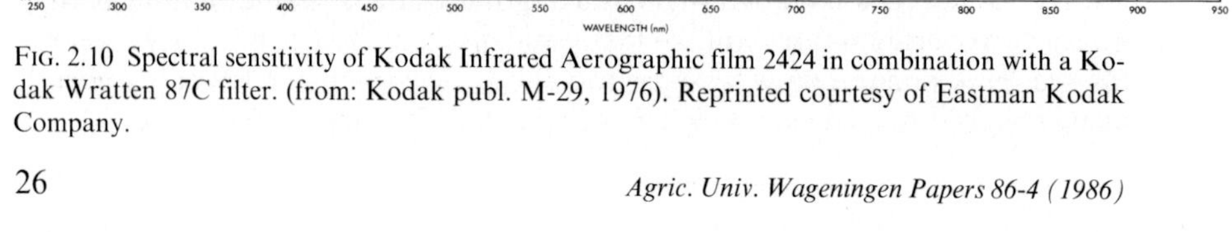


FIG. 2.10 Spectral sensitivity of Kodak Infrared Aerographic film 2424 in combination with a Kodak Wratten 87C filter. (from: Kodak publ. M-29, 1976). Reprinted courtesy of Eastman Kodak Company.

TABLE 2.3 Comparison between a hand-held radiometer, a field spectroradiometer, aerial multispectral scanning (MSS) and aerial multispectral photography (MSP)

| | hand-held radiometer | field spectroradiometer | MSS | MSP |
|---|---|---|---|---|
| investments[1] | high | high | very high | high |
| exploitation costs[1] | low | low | high | moderate |
| data registration | digital | digital | digital | analog |
| availability data | immediately | few days | few weeks | few days |
| spatial resolution[2] | 0.5 $m^2$ | 1 $m^2$ | 1–4 $m^2$ | 0.1–1 $m^2$ |
| spectral resolution[2] | 10–100 nm | 10–50 nm | 50–100 nm | 25–100 nm |
| measurable area[3] | 100–300 $m^2$ | 20–50 $m^2$ | > 100 ha | > 100 ha |

[1] Order of magnitude: low = less than f 100
moderate = f 100–1000
high = f 10000–50000
very high = more than f 50000.
[2] Highest resolution of available equipment.
[3] Order of magnitude measurable within a few hours (2–5 hours) with the available equipment.

Black and white aerial photography was found to be preferable because of the more straightforward separation of bands with a one-layer film. The spectral resolution may be high when adequate films and filters are used in order to achieve a multispectral photographic (MSP) system with narrow bands (figure 2.10). Consequently, cost stays within acceptable limits. Yet this system can be applied to relatively large areas (several hectares). All these considerations render black and white multispectral aerial photography the most promising remote sensing technique for application to field trials. The only limitation is that calibration and use of the sensor system must be very accurate if this technique is to supply quantitative information, and this has often been a bottle-neck for its application (e.g. Sievers, 1976). In this thesis, procedures for solving these problems will be given (chapter 3). The system was also tested at the ir. A.P. Minderhoudhoeve.

All four systems are summarized in table 2.3.

## 2.5 SELECTING THE OPTIMAL CHANNELS

A question that may arise is whether all channels in the optical region mentioned (visible and reflective infrared) are necessary for estimating crop characteristics. Ideally, the maximum amount of information should be obtained from these spectral measurements but the total number of channels needed should be limited, as this could lead to savings in costs of data collection, processing (by computer) and interpretation.

The choice quoted from literature in section 2.3 was verified by using results from the MSS recordings.

*2.5.1 Selection criteria*

The MSS data were statistically analysed so that a subset of multispectral channels containing the smallest set of regressors and which explained most of the variability in the response variable could be selected. Yield was used as the response variable. The analysis was restricted to linear multiple regression, i.e. with equations that are linear in their unknown constants:

$$\text{yield} = a_0 + a_1\,\text{Ch5} + a_2\,\text{Ch6} + a_3\,\text{Ch7} + a_4\,\text{Ch8} + a_5\,\text{Ch9} \qquad (2.1)$$

Linear regression equations between spectral measurements and crop characteristics have been reported in many investigations (e.g. Barnett & Thompson, 1983; Holben et al., 1980; Pollock & Kanemasu, 1979; Tucker et al., 1980). Pearson & Miller (1973a) and Verhoef (1979) also applied linear equations for selecting an optimal subset of channels. In the literature some of the relationships appeared curvilinear in graphic presentation, and therefore quadratic terms were also incorporated in the linear model. Under the conditions of the present study, coefficients of quadratic terms were not significant. This was also found by Ahlrichs & Bauer (1983). The range in the yield data used was mostly not very large (cf. coefficients of variation for the yield data in the following sections). So, regression curves of yield on reflectance measures were linear for these data.

Since ratios have often been used as indices for estimating crop characteristics (e.g. Holben et al., 1980; Pearson et al., 1976; Tucker, 1979; Tucker et al., 1980), a logarithmic transformation of the radiances in the distinct channels was also applied in this study. This transformation did not yield better models, so only linear models that are linear in the regressors were used for selecting the optimal subset of channels (cf. equation 2.1). This subset selection is the only aim of the rest of this chapter, and therefore little attention will be paid here to the explanation of the relationships found. This will be given more emphasis in later chapters.

To establish an optimum configuration, all models consisting of all possible combinations of present channels have to be compared. Many linear models may be possible. The 'Linwood' computer program (Daniel & Wood, 1980) was used to compare all possible models. The MSS channels (radiance values) are the regressors used for fitting to a crop characteristic. Using a $C_p$ value, the model giving a good fit to the response variable (crop characteristic) and consisting of as few regressors as possible is selected. This $C_p$ value, sensitive to bias (= lack of fit) because of the omission of relevant regressors, is defined as (according to Daniel & Wood, 1980):

$$C_p = \frac{RSS_p}{s^2} - (N-2p) \qquad (2.2)$$

$RSS_p$ = residual sum of squares for the model with p regression coefficients $(= \Sigma_i(y_i-\hat{y}_i)^2)$

$s^2$ = estimated variance of the observations

N = total number of observations
p = number of regression coefficients to be estimated.

$s^2$ is often estimated from the full model (with all regressors present), but is in fact the true variance ($\sigma^2$) or its estimate for a correct model (the model with all relevant regressors, whether measured or not, necessary for explaining the response variable). This may result in the $C_p$ value being underestimated, which is undesirable.

As an alternative it is preferable to look at the residual sum of squares ($RSS_p$) for each model (Jansen, 1978). Plotting these residual sums of squares against the corresponding number of degrees of freedom (N-p) may result in a picture similar to that shown in figure 2.11. Points in the model whose connection line with the origin have smaller slopes but which have large (N-p) values, are candidates for selection. The slope of the line connecting points with small and more or less constant slopes represents a valid estimate of the variance of the observations.

In order to make different experiments more readily comparable we use the (estimated) coefficient of variation (CV):

$$CV = \frac{\sqrt{RSS_p/(N-p)}}{\bar{y}} \tag{2.3}$$

$\bar{y}$ = mean value of the dependent variable.

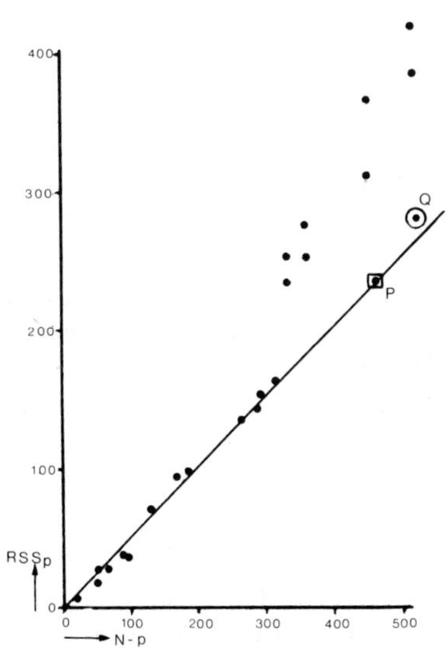


FIG. 2.11 Example of the residual sum of squares as a function of the corresponding number of degrees of freedom (After: Jansen, 1978). P or Q are adequate subsets.

The optimal model selected from figure 2.11 will be the model with minimum CV.

In this study, selecting the optimal subset of channels based on the $C_p$ value and that on the CV value both yielded identical results with the data sets concerned.

Recordings of multispectral scanning were obtained at the ir. A.P. Minderhoudhoeve on 8 July 1981 and 13 July 1982. The azimuth angle between sun and sensor was about 90° or 270° (scan lines oriented perpendicular to the azimuth of the sun), because at such angles the predicted off-nadir viewing effects are minimal (Kimes, 1983; Slater & Jackson, 1982). Several field trials were analysed. The mean of all net pixels per plot (10-15) was calculated and digital pixel values were converted into radiance ($mWm^{-2}sr^{-1}$*).

*2.5.2 Field trial 105 in 1981 (wheat)*

At the ir. A.P. Minderhoudhoeve a uniformity trial was laid out in 1981 with spring wheat, cultivar Bastion. The aim was to investigate whether systematic differences between plots occurred as a result of local differences in soil fertility. The trial consisted of 144 uniformly fertilized plots. The wheat was sown on 6 April 1981, and harvested on 31 August 1981. Grain yield only was ascertained (at the end of the season). Mean yield for these 144 plots was 6 300 kg/ha and the coefficient of variation (CV) was small: 0.035. Because of this small variation among the yield data the explanation of yield differences by means of MSS data was not expected to be spectacular. Yet, an analysis along these lines was carried out.

An MSS mission was carried out on 8 July 1981. At that moment the ears of the wheat crop were just visible and the crop was still completely green (Feekes stage 10.5, cf. appendix 1).

2.5.2.1 Optimal choice of channels

Because digital pixel values and radiance values are linearly related it was immaterial whether the analysis was carried out on the basis of the one or the other. Radiance values were used in the computations.

Some of the channels correlated strongly with one another. Channels 5 and 6 (correlation coefficient $r = 0.92$) and channels 7 and 8 ($r = 0.94$) correlated strongly. Channels 8 and 9 correlated less strongly ($r = 0.80$).

A linear model with channels 5, 7 and 9 seemed to offer a good fit of yield with $CV = 0.025$. This model is given in equation (2.4):

$$\text{yield} = 3550 - 2.07Ch5 - 2.05Ch7 + 1.43Ch9 \qquad (2.4)$$

Study of the MSS images showed that the radiance in some channels may depend on the view angle; this may have influenced the above analysis. The importance of this angle will be examined further in the following section.

* sr = steradian

2.5.2.2 View angle effects

The view angle effects are illustrated in figure 2.12. Field trial 105 consisted of 12 strips of 12 plots each, such that each strip was situated parallel to the flight direction (cf. figure 2.8). Thus for each individual strip the view angle was constant. View angle 0° (vertically downwards) occurred about half-way between strips 3 and 4. The optimal subset of channels was also selected for each individual strip.

As compared with the model concerning all six channels, the model with channels 5, 7 and 9 offered the optimal fit of grain yield (table 2.4) for each individual strip.

Two solutions for taking the view angle into account were applied.

1. The first method is to introduce the view angle into the model as an additional regressor. Because the view angle effect may differ for the different channels, the interaction terms between view angle and the channels were also included in the calculations. The introduction of the view angle effect appeared to yield some improvement of the results and the optimal model contained the view angle, channels 5 and 9 and the corresponding interaction terms (CV = 0.022).

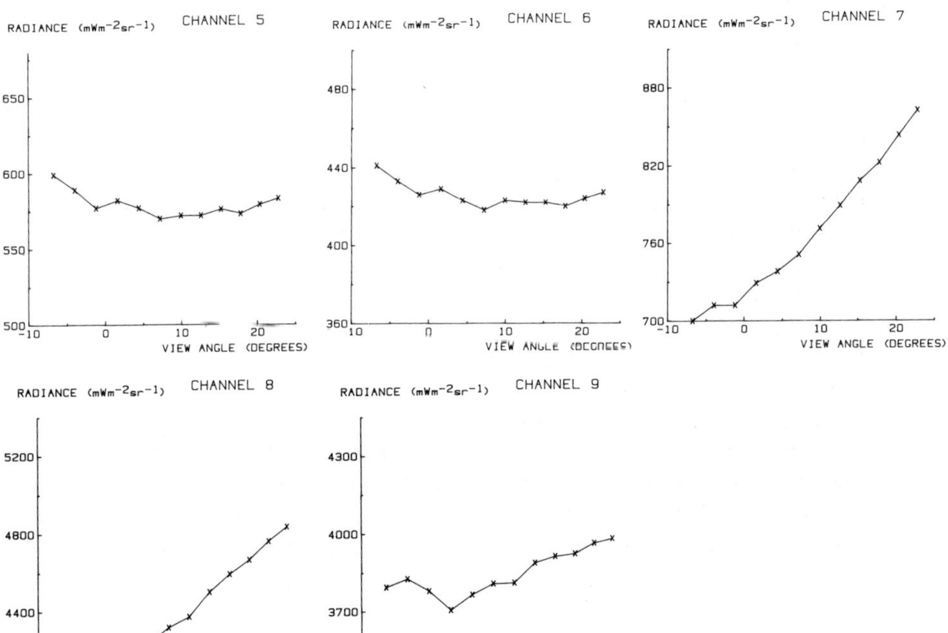


FIG. 2.12 View angle effects with the MSS channels in field trial 105. Recording date: 8 July 1981 (azimuth angle between sun and flight direction: 0°).

TABLE 2.4 Coefficients of variation of residuals when estimating grain yield of field trial 105 from MSS data for the linear model with channels 5, 7 and 9 as compared with the model incorporating all 5 channels for each individual strip.

| strip | view angle (°) | channels 5, 6, 7, 8, 9 CV | channels 5, 7 and 9 CV |
|---|---|---|---|
| 1 | 6.8 | 0.020 | 0.019 |
| 2 | 4.0 | 0.016 | 0.014 |
| 3 | 1.2 | 0.024 | 0.025 |
| 4 | 1.6 | 0.015 | 0.015 |
| 5 | 4.4 | 0.016 | 0.018 |
| 6 | 7.2 | 0.026 | 0.027 |
| 7 | 9.9 | 0.021 | 0.021 |
| 8 | 12.6 | 0.023 | 0.021 |
| 9 | 15.2 | 0.023 | 0.020 |
| 10 | 17.8 | 0.028 | 0.029 |
| 11 | 20.3 | 0.015 | 0.018 |
| 12 | 22.7 | 0.022 | 0.026 |

2. The second method involves correction for the gradient of the measured radiance, perpendicular to the flight direction. A second order polynomial correction for a gradient in the scan direction (x-direction) was carried out on the image that covered field trial 105. This function appeared to provide a good fit to the observed gradient (a second order polynomial function was also found by Ott et al., 1984). After this correction, the model that used all 5 channels to fit grain yield resulted in CV = 0.024. These results were no better than those obtained without the correction for the gradient and were worse than those obtained when the view angle had been introduced into the model. Also, selecting the optimal submodel with correction offered no improvement over the submodels without correction for a density gradient in the scan direction.

Thus equation (2.4) was judged to be the optimal submodel.

### *2.5.3 Field trial 92 in 1981 (wheat)*

During the 1980/1981 season a field trial was carried out at the ir. A.P. Minderhoudhoeve for investigating the effects of plant density, nitrogen nutrition and fungicide treatment on grain yield with winter wheat (cultivar Arminda) and spring wheat (cultivar Bastion). Winter wheat was sown on 14 October 1980 and spring wheat on 2 April 1981. Treatments were 2 plant densities (125 and 250 seeds per $m^2$), 2 fungicide treatments (3 kg Bavistin M per ha or none) and 3 nitrogen applications as follows:

| Winter wheat: | | | | | |
|---|---|---|---|---|---|
| Feekes stage | F4 | F6 | F8 | F10 | total |
| N1 | 60 | – | – | 60 | 120 kg N per ha |
| N2 | – | 60 | – | 60 | 120 kg N per ha |
| N3 | – | – | 60 | 60 | 120 kg N per ha |

| Spring wheat: | | | | | |
|---|---|---|---|---|---|
| Feekes stage | F0 | F6 | F8 | F10 | total |
| N1 | 30 | – | – | 55 | 85 kg N per ha |
| N2 | – | 30 | – | 55 | 85 kg N per ha |
| N3 | – | – | 30 | 55 | 85 kg N per ha |

The whole trial was designed in three replicates. Each net plot size was 3 m by 17 m. On 18 August 1981 the winter wheat was harvested and on 1 September 1981 the spring wheat was harvested. On 8 July 1981 an MSS mission was carried out as described in section 2.5.1.

The optimal subset of channels for fitting grain yield was ascertained. Because spring wheat and winter wheat were different crops with different growth stages at the recording date, data on each crop were analysed separately.

2.5.3.1 Optimal choice of channels for spring wheat

Because of the introduction of treatment effects, the variation within the yield data was larger than that in field trial 105. In the analysis one plot appeared to be an outlier and was omitted in the subsequent analysis. The mean yield for the remaining 35 plots with spring wheat was 5300 kg/ha, with a CV value of 0.127.

On the day of the MSS mission the spring wheat still had a green canopy. The ears had appeared and the crop was at the flowering stage (Feekes stage 10.5: F10.5). MSS channels 5 and 6 ($r = 0.88$) and channels 8 and 9 ($r = 0.87$) correlated strongly. The optimal channels for the linear model were channels 5, 7 and 9 ($CV = 0.086$). This model is given in equation (2.5):

$$\text{yield} = -4150 - 15.3\text{Ch5} + 9.51\text{Ch7} + 3.08\text{Ch9} \qquad (2.5)$$

Field trial 92 consisted of 6 strips parallel to the flying direction. Each strip contained 6 plots of spring wheat (and 6 plots of winter wheat). View angle $0°$ occurred somewhere between strips 3 and 4 and the maximum view angle was about $7°$. The results from field trial 105 indicated that up to such a view angle, differences in reflectance resulting from this changing view angle would be small. This was also found by other researchers, e.g. Barnsley (1984), Kimes (1983), Koepke & Kriebel (1978), Malila (1968), Ott et al. (1984), Slater & Jackson (1982). Indeed, the introduction of the view angle as an additional regressor (also including interaction terms between view angle and the various channels) did not improve the fit of grain yield.

#### 2.5.3.2 Optimal choice of channels for winter wheat

The MSS mission for the winter wheat crop was performed towards the end of the growing season. The grains were already maturing and senescence of the canopy was proceeding (F11). So the canopy was no longer green. The effects of senescence may disturb the fitting of grain yield to spectral measurements. The mean yield for all 36 plots with winter wheat was 7 400 kg/ha, with a CV value of 0.131.

Again, channels 5 and 6 (r = 0.93) and channels 8 and 9 (r = 0.97) correlated strongly. The optimal selection of channels contained nearly all channels (channels 6, 7, 8, 9) with CV = 0.108. The introduction of the view angle resulted in a minor improvement of the model. The moment chosen for the recording, i.e. at the end of the growing season, was probably a bad choice for explaining differences in yield.

### *2.5.4 Field trial 116 in 1982 (barley)*

Field trial 116 was designed for investigating the influence of sowing date and nitrogen nutrition on grain yield with barley, cultivar Trumpf. Treatments were 2 sowing dates, 26 March and 22 April, and 6 nitrogen levels (0–20–40–60–80–100 kg N per ha). The trial was designed in 3 replicates. The size of each net plot was 3 m by 15 m. On 17 August the barley was harvested by combine. On 13 July an MSS mission was carried out as described in section 2.4, but now also with channel 4 (500–550 nm) included, to ascertain whether this channel offers additional information.

The two sowing dates meant that there were two different growth stages on 13 July with different reflectance patterns. These were analysed separately.

#### 2.5.4.1 Optimal choice of channels

Channels 4, 5, 6 and 7 correlated strongly with each other. Their mutual correlation coefficients were at least 0.99. Channels 8 and 9 also correlated strongly (r = 0.90).

In the early-sown crop, senescence had already started (F11) on the date of the mission (13 July 1982). The mean grain yield of all 18 plots sown early was 7 100 kg/ha with a CV value of 0.109. The optimal subset for the early-sown crop contained channels 6, 7, 8 and 9 (CV = 0.080), although for the whole field trial channels 6 and 7 and channels 8 and 9 correlated strongly. As with field trial 92 in 1981, only moderate results were obtained in explaining differences in grain yield from spectral measurements at senscence.

In the late-sown crop the canopy was still green, although the ears had appeared (F10.5) by 13 July. For this sowing date the mean grain yield was 6 500 kg/ha with a CV value of 0.053, which shows the variability within these data to be small. The optimal subset now contained channels 5, 7 and 9 (CV = 0.025). This model was satisfactory for explaining grain yield. Because this field trial consisted of 3 strips of plots only, the differences in view angle were very small. Thus the view angle was not taken into account. This model is given in equation (2.6):

$$yield = -341 - 7.97Ch5 + 7.22Ch7 + 1.36Ch9 \quad (2.6)$$

## 2.6 Summary

Remote sensing recordings in the visible and reflective infrared regions of the electromagnetic spectrum can offer characteristic spectral information, which may be related to crop characteristics such as soil cover, leaf area index or dry matter weight. From an aerial platform, recordings in the green (at 550 nm), in the red (at 670 nm) and in the infrared (at 870 nm) are optimal (derived from Bunnik, 1978).

The results of our tests showed that an aerial platform is most appropriate for recording extended field trials. Multispectral scanning (MSS) is too expensive for very frequent recordings during the growing season. By using 70-mm aerial photography the cost should stay within acceptable limits, thus offering a high temporal resolution. If the films and filters are chosen carefully, a high spectral resolution should be obtained. By adjusting the aperture of the densitometer to variables such as lens type, flight altitude and image motion, a high spatial resolution should also be obtained. This renders black and white multispectral aerial photography the most promising remote sensing technique for application to agricultural field trials. The specifications and calibration of this system will be described in the next chapter.

An analysis of MSS data showed that channels 4, 5 and 6 of the scanner used always correlated strongly, as did channels 8 and 9. The optimal selections of channels mostly contained the channels 5 (550–600 nm), 7 (650–700 nm) and 9 (800–890 nm). These results confirmed the choice obtained from the literature (e.g. Bunnik, 1978).

# 3 DESCRIPTION OF MULTISPECTRAL AERIAL PHOTOGRAPHY

## 3.1 Introduction

Chapter 2 resulted in the conclusion that the most appropriate technique for frequent acquirement of remotely sensed data about field trials at the ir. A.P. Minderhoudhoeve is black and white multispectral aerial photography, recordings being obtained in the visible and (reflective) infrared region of the electromagnetic spectrum. By using available camera equipment the flight height for recording has to be adjusted to the size of the area (field trials) to be covered by one photograph. To obtain the optimal passbands (section 2.3) with high spectral resolution as well as a satisfactory radiometric resolution (which is also influenced by film processing), the films and filters used have to be carefully selected. After satisfying the above conditions an appropriate densitometer has to be selected and its aperture has to be adjusted in order to obtain a satisfactory spatial resolution for application at field trials.

The author and Mr. C. Horton of the Polytechnic of Central London collaborated to study ways in which photographic recordings could be calibrated. In particular the calibration concerns the characteristic curve, light fall-off, exposure time, relative aperture, focal setting and atmospheric correction. The required experiments and the development of algorithms for radiometric calibration were carried out within the research programme for this thesis. The theoretical and physical background of the various photographic problems is being studied extensively at the Polytechnic of Central London by C. Horton. In the following sections only the practical solutions for these problems will be discussed.

## 3.2 Platform and camera equipment

During the 1982 and 1983 growing seasons the aerial platform was a single-engine aircraft (Piper Archer) with ports allowing a navigation sight and two anti-vibration camera mounts with drift setting and intervalometer control. Two Hasselblad cameras equipped with Zeiss Planar 100 mm lenses (or Zeiss Distagon 50 mm lenses) could operate in the aeroplane to obtain simultaneous recordings on two film materials (figure 3.1). Since 70-mm film material was used, the effective size of the negative was about 55 mm by 55 mm. In appendix 2 a plan of the experimental farm, where the research was carried out, is shown. It was desirable for one vertical photograph to cover an area at least 300 metres long–which was the maximum length of a field trial. On successive flights over the area, adjacent strips should have sufficient sidelap (about 30%) in order to ensure total coverage. So, the scale should be about 1 : 8 000. With a 100 mm lens (wich is the most suitable lens of those available at the Agricultural

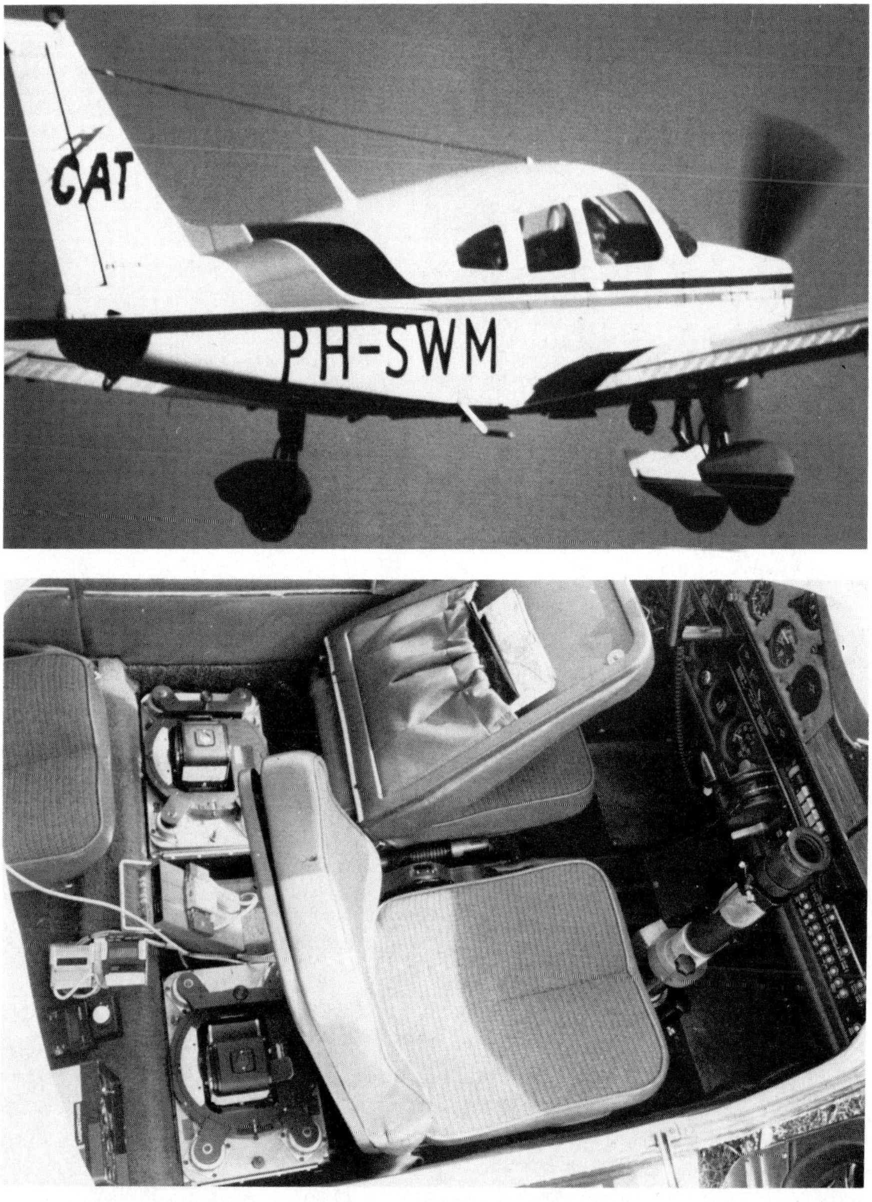


FIG. 3.1 The single-engine aircraft with camera mounts and cameras for 70-mm films.

University) this resulted in a flight height of 800 metres, while for a 50 mm lens (used when the sky was completely overcast, with a low cloud base) this resulted in a flight height of 400 metres. Photographs were taken at intervals of about two weeks to keep in step with conventional field sampling. Flights

FIG. 3.2 The microlight.

were performed at a mean solar altitude of 45 degrees, after it had been ascertained that weather conditions were constant and cloud shadow was absent at the site.

Cloud shadow at the recording site can pose huge problems with aerial photography in general. If one has to extract quantitative information out of such recordings one will look at (or measure) tonal differences within the image. However, cloud shadow will also introduce tonal differences, which are nearly impossible to separate accurately from tonal differences caused by different objects. The weather in the Netherlands may change rapidly, especially with respect to cloud cover. With the Piper Archer platform it takes several hours between the moment that it is decided to execute a mission and the moment that the photography can start. The crew (pilot, navigator and photographer) have to travel to the airfield with the equipment (1 hour), they have to install all the equipment in the plane (three-quarters of an hour) and they have to fly to the recording site (twenty minutes). Meanwhile the cloud cover may have changed dramatically.

In order to have a platform more quickly available at the recording site, the experimental farm of the Agricultural University bought a small microlight aircraft, the American Aerolights' 'Eagle'. On this platform an anti-vibration camera mount for two Hasselblad cameras was also attached (figure 3.2). A drift setting and intervalometer control were also installed. Much experience was obtained with the microlight during the 1982 and 1983 growing seasons. The microlight has been operational since the 1984 season. Vertical photographs have been taken at an altitude of 300 metres (scale 1 : 3 000). So, often more than one photograph is required to cover a whole field trial. Personnel at the experimental farm fly the microlight and handle all the equipment. It has proved possible to start photography within 10 to 15 minutes after weather conditions have been judged to be favourable. Results obtained with the microlight as a platform will not be given in this monograph (which only deals with trials from 1981 to 1983), but will be published in other reports.

## 3.3 Choice of film and filters

Kodak publications M-29 and B-3 were studied to ascertain the best aerial films with high sensitivity that should be used in combination with filters to match the spectral passbands that our research described in the previous chapter had shown to be optimal for the Daedalus scanner. The aim was to facilitate

TABLE 3.1 Film and filter combinations used in the present study to match the optimal spectral bands.

| Band | Kodak 70-mm Aerial film | Kodak Wratten filters | wavelength (50% sensitivity) | bandwidth |
|---|---|---|---|---|
| green | PX 2402 | W21 + W57A | 555–580 nm | 25 nm |
| red | PX 2402 | W70 | 665–700 nm | 35 nm |
| infrared | IR 2424 | W87C | 840–900 nm | 60 nm |

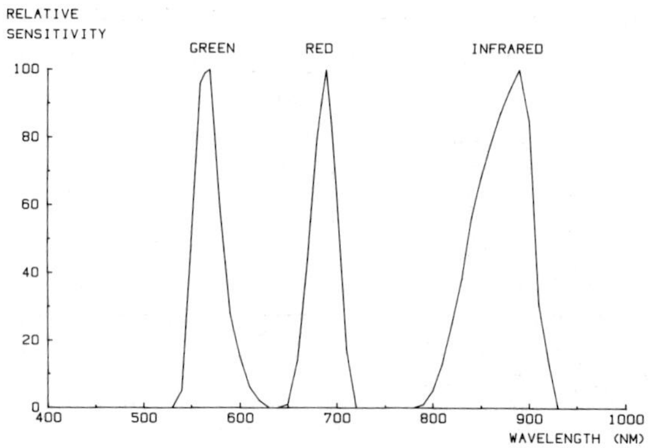


FIG. 3.3 Spectral sensitivity of the green, red and infrared passbands used with multispectral aerial photography.

the comparison of the results obtained with aerial photography with those obtained with multispectral scanning. The film/filter combinations used are given in table 3.1. The relative spectral sensitivity (determined as a combination of film sensitivity and filter transmittance) of these three passbands is illustrated in figure 3.3.

## 3.4 FILM EXPOSURE

First, an experiment was designed to ascertain the correct exposure with the various film/filter combinations. Exposures of Kodak greycharts were made at ground level. A Minolta photometer (type Flash Meter III) was used so that the camera exposure control settings could be adjusted for variations in illumination. The films were processed in a standardized manner as described in section 3.7. For a correct exposure the density of the darker Kodak greychart (about 18% reflectance) should be about 1.0. The camera settings for correct exposure were determined by the effective aerial film speed (E.A.F.S.) given in table 3.2.

However, if the object is a natural surface, or, more specifically, a vegetation, the reflectance in the various passbands will not be 18%. From literature and from results reported in this monograph it may be deduced that with vegetation the reflectance in the visible region of the spectrum can be considerably lower than 18%. Reflectances of 5% are quite common, and in the red region even much smaller values are possible. In the infrared region reflectance values up to 50% or 60% are commonly encountered. One has to consider the whole range of possible reflectance values, also taking into account that the atmosphere causes some additional contribution to aerial measurements. The latter implies that if an object reflects e.g. 5% of the incoming radiation, the reflectance at sensor level may be e.g. 10% because of the addition of sky light.

Thus, a camera setting adjusted to an E.A.F.S. of 12 ASA in the visible pass-

TABLE 3.2 Effective aerial film speed (E.A.F.S.) in ASA for adjusting camera settings in order to obtain correct exposures with various film/filter combinations.

| film* | E.A.F.S. without filter | filters | E.A.F.S. with filters | E.A.F.S. applied during missions |
|---|---|---|---|---|
| PX 2402 | 160 | W21 + W57A | 25 | 12 |
| PX 2402 | 160 | W70 | 35 | 12 |
| IR 2424 | 320 | W87C | 35 | 50 |

* PX 2402 was developed with DK-50 for 18 minutes at 20°C.
IR 2424 was developed with DK-50 for 16 minutes at 20°C.

bands and to an E.A.F.S. of 50 ASA in the infrared passband should result in an adequate exposure of objects that range from bare soil (wet or dry) up to vegetation with a high LAI. All density values should, as far as possible, be in the 'linear' part of the D-logQ curve (section 3.7); in other words, they should be in the density range of 0.5 to 2.0.

In order to verify the above exposure values a mission was carried out in the beginning of 1982 with the film/filter combinations described in table 3.1. At the test site, reference targets (see section 3.12) were placed in order to obtain objects early in the growing season that covered the whole range of possible reflectance values. In practice, camera settings according to the above E.A.F.S. values satisfied the conditions made for usable density values, and these E.A.F.S. values were used during all missions for this research.

## 3.5 DENSITOMETRY

Exposing a point on a film results in the film darkening because silver is deposited on that point during development. The darkening of the film is usually expressed as a density (D), which is defined as:

$$D = \log(1/T) \qquad (3.1)$$

with T = transmittance of the film (expressed as a fraction).

The relationship between energy and density is described by the characteristic curve (section 3.7). Densities are measured with a densitometer.

At the Wageningen Agricultural University a Macbeth TD-504 densitometer was available. Because the plots in the field trials studied were very small (plot width of 3 metres), the aperture of the densitometer also had to be small. With a scale of 1 : 8 000 a plot width of 3 metres corresponds to a distance of 0.375 mm on the negative. The longest exposure time during recording was 1/60 seconds; the flight speed of the Piper Archer was about 50 metres per second. This resulted in a maximum displacement during one exposure of less than 1 metre, corresponding with an image motion of less than 0.1 mm at a scale of 1 : 8 000. Therefore a special aperture with a diameter of 0.25 mm was constructed and substituted for the original apertures of 1 mm, 2 mm and 3 mm. As an alternative to the Macbeth densitometer it was also possible to use a Joyce-Loebl 3 CS microdensitometer at the Polytechnic of Central London. With the microdensitometer it is possible to use very small apertures (the aperture may vary from 50 × 1 μm to 6 × 1 mm).

An experiment was conducted to compare results obtained with the Macbeth densitometer, aperture 0.25 mm, with the results obtained with the Joyce-Loebl microdensitometer, aperture 0.033 × 0.066 mm. For this experiment, recordings from 11 August 1982 were used. With both instruments a scan (i.e. density recordings in one direction) was made across an object, and the result was recorded on an analogue recorder. The reference targets set up along a line, and field trial 116, were used as objects. Field trial 116 consisted of 3 parallel strips with 12 plots each (a total of 36 plots). Only one scan was made across each strip, whilst it was not possible to determine the position of this scan very accurately. This implies that the identical scan line across a strip was not measured with both instruments, although the objects were the same. The recordings were obtained as described in the previous sections.

Figure 3.4 illustrates the result of a scan across the reference targets with the microdensitometer and the densitometer, respectively. The scaling in the scan direction differs because of a difference in scan speed and speed of the recorder. The pattern is essentially the same, although with the microdensitometer the curve is less smooth (more 'noise') because of the small aperture. The results

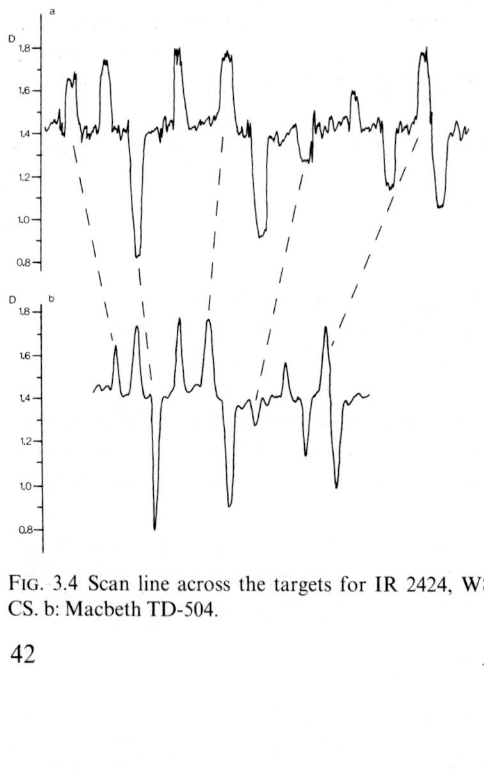


FIG. 3.4 Scan line across the targets for IR 2424, W87C, on 11 August 1982. a: Joyce-Loebl 3 CS. b: Macbeth TD-504.

TABLE 3.3 Comparison of density measurements of the targets with the Joyce-Loebl 3 CS microdensitometer (= J-L) and with the Macbeth TD-504 densitometer (= M). Field trial 116, 11 August 1982.

| target | PX 2402 | W21+W57A | PX 2402 | W70 | IR 2424 | W87C |
|---|---|---|---|---|---|---|
| | J-L | M | J-L | M | J-L | M |
| 1 grey | 2.20 | 2.08 | 2.25 | 2.23 | 1.75 | 1.74 |
| 2 blue | 1.15 | 1.15 | 1.20 | 1.20 | 1.13 | 1.14 |
| 3 green/blue | 1.41 | 1.40 | 1.37 | 1.37 | 1.56 | 1.55 |
| 4 10% | 1.52 | 1.53 | 1.58 | 1.58 | 0.90 | 0.90 |
| 5 25% | 1.94 | 1.93 | 2.02 | 2.02 | 1.25 | 1.28 |
| 6 60% | 2.22 | 2.22 | 2.31 | 2.31 | 1.74 | 1.76 |
| 7 white | 2.21 | 2.19 | 2.30 | 2.30 | 1.77 | 1.77 |
| 8 black | 1.11 | 1.12 | 1.18 | 1.17 | 0.80 | 0.79 |
| 9 red | 1.47 | 1.48 | 2.09 | 2.08 | 1.71 | 1.72 |
| 10 green | 1.53 | 1.54 | 1.56 | 1.55 | 1.63 | 1.63 |

of the density measurements of the targets in the three passbands described in section 3.3 are given in table 3.3, an average value per target being used for the measurements with the microdensitometer. The results are very similar. An example of a comparison of scan lines across one strip of field trial 116 is given in figure 3.5. Figure 3.6 presents the densities measured in the 36 plots of field trial 116 with both instruments in all 3 passbands. The density values obtained are the result of only one scan across each plot, and therefore small discrepancies will appear because of natural variation within a plot. For convenience, dashed lines indicate a discrepancy of 0.03 density between results from both instruments (it was assumed that measurements with both instruments were carried out with a standard deviation of about 0.02 density: therefore a discrepancy of about 0.03 density was expected).

From the results it was concluded that a densitometer (in this case the Mac-

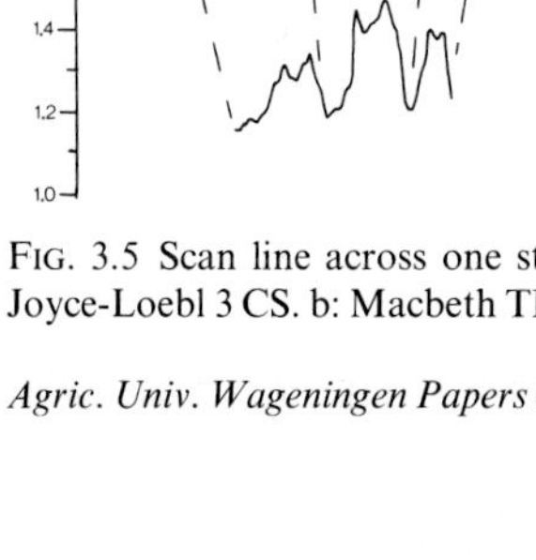


FIG. 3.5 Scan line across one strip of field trial 116 for IR 2424, W87C, on 11 August 1982. a: Joyce-Loebl 3 CS. b: Macbeth TD-504.

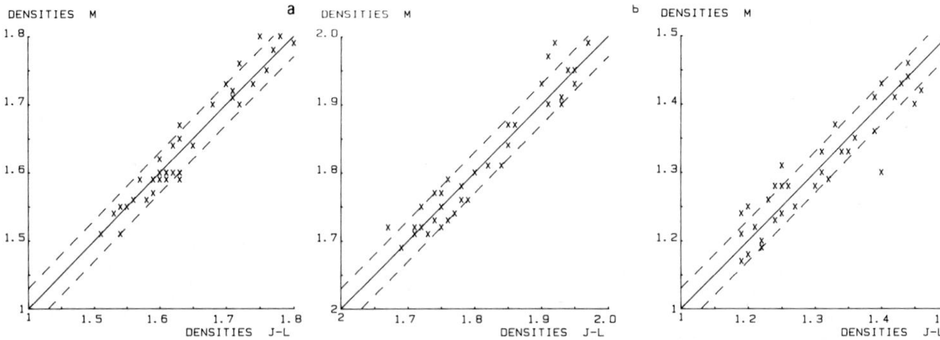


FIG. 3.6 Comparison of density measurements in field trial 116 on 11 August 1982 with the Joyce-Loebl 3 CS (J-L) and the Macbeth TD-504 (M). —: M = J-L --- : M = J-L ± 0.03. a: PX W21 + W57A (green). b: PX W70 (red). c: IR W87C (infrared).

beth TD-504) with an aperture of 0.25 mm diameter is suitable for analysing 1 : 8 000 scale photographic recordings of field trials with plots of 3 metres net width. A much smaller scale will pose problems because the images of the plots will be smaller than the aperture. By using a microdensitometer one is measuring much variation (and noise) within a plot largely caused by the granularity of the image (less so by differences in reflectance within one plot). In field trials one is not interested in the variation within the various plots, but only in an average value per plot. These results show that it is even easier to extract such an average by using a densitometer, because of its smoothing effect. Yet, extreme values (e.g. of reference targets) of such averages will not be removed by smoothing if the aperture is not too large. It was decided to use the Macbeth TD-504 densitometer to analyse all the images obtained in this study.

Another problem was to ascertain the number of scan lines that have to be measured across each strip of plots to obtain a correct mean value per plot. In order to investigate this, one negative of field trial 116 in 1982 was analysed (IR 2424, W87C, 27 May 1982). Net plot size was 3 m by 17 m. Plots of this size were common on the ir. A.P. Minderhoudhoeve. Between 1 and 9 equidistant scan lines were measured per strip of 12 plots, and per scan line the mean reflectance was calculated for each plot. Subsequently, the mean value per plot was calculated by using different numbers of scan lines, ranging from 1 to 9. As a measure of precision the coefficient of variation (CV) of residuals, resulting from the analysis of variance (cf. chapter 6), was used to ascertain the optimal number of scan lines per strip. The results are illustrated in figure 3.7. The CV value should be as small as possible, but also the number of scan lines needed should be kept minimal. It was decided that three scan lines across each plot would be optimal for this investigation (CV = 0.03).

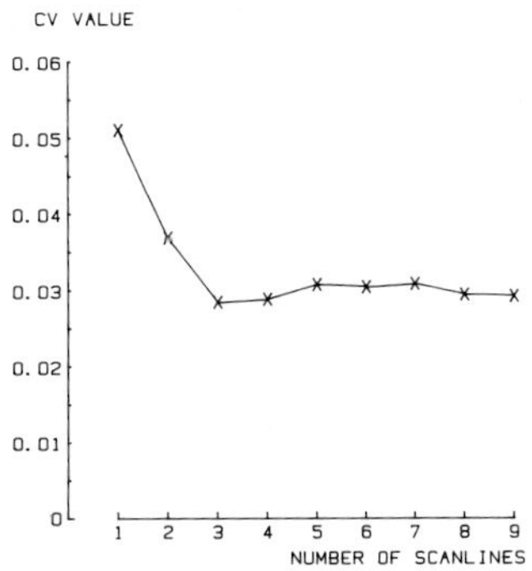


FIG. 3.7 Coefficient of variation (CV) as a function of the number of scan lines per plot. Field trial 116, IR 2424, W87C, 27 May 1982.

## 3.6 DENSITOMETRY AND PHOTOGRAMMETRY

A problem inherent to data extraction from multidate recordings is caused by differences in image location and scale between object points as a result of slight differences in flight paths and altitudes. However, the aim is to be able to measure across the same part of an object or plot repeatedly. Therefore, prior to the flying missions a rectified photomap of the trial area was made from photographs scale 1 : 3 000 taken by a 230 mm × 230 mm metric camera. Then the image coordinates of pass points (in the terrain) were accurately determined by using a stereo analogue plotter, and were stored in the memory of an HP-85 microcomputer. Subsequently, the positions of scan lines relative to these pass points were determined and also stored in the memory of the microcomputer. This facilitated precise scanning regimes to be made of selected plots with a densitometer.

The Macbeth TD-504 densitometer, aperture 0.25 mm, was modified so that it would operate as a flat-bed scanning densitometer (figure 3.8). A frame was built to support the film. This frame was driven in the x and y directions by stepping motors, and it could be rotated manually. The position of the frame (and thus of any point of the image it supports) was digitally stored by means of a digitizer tablet (accuracy ± 0.05 mm) which was interfaced with an HP-85 microcomputer. The densitometer itself was also interfaced with the microcomputer for density readings (in units of 0.01 density). First, pass points were digitized, followed by frame corner points. With the microcomputer the image coordinates were transformed to terrain coordinates in the system derived from the rectified photomap. The screen of the microcomputer displayed how much the frame should be rotated to obtain the correct scan direction, and for each scan line it displayed the displacement in x and y to position the frame at the desired starting point. After an image was correctly positioned, readings were made at discrete intervals (16 $mm^{-1}$) in the x direction and densities were recorded together with corresponding photo coordinates. The fact that there were 16 read-

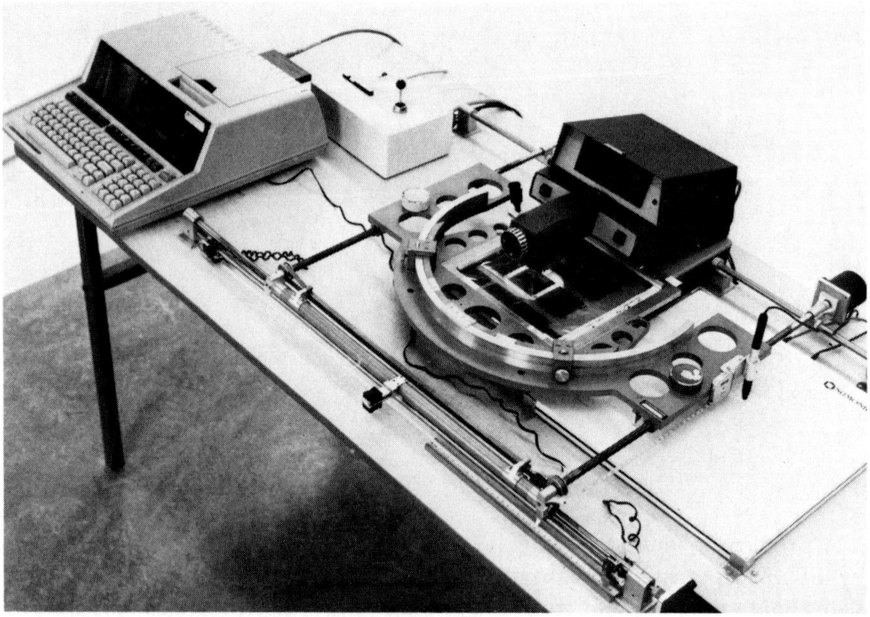

FIG. 3.8 Macbeth TD-504 densitometer modified for controlled line-scanning.

ings per mm implied that there was considerable overlap between consecutive measurements. This was necessary if at least one density reading was to be obtained per 3 m-wide net plot, without interference from the edges.

It was also possible to interface the microcomputer with the main computer (DEC10) of the Agricultural University. With the main computer some correction algorithms were applied to the measurements, in order to correct for factors described in the following sections.

## 3.7 CHARACTERISTIC CURVE (SENSITOMETRY)

The characteristic curve of a processed film is a plot of its optical densities against the logarithm of the corresponding exposures. The shape of this curve for black and white film materials is illustrated in figure 3.9. The density level $D_{min}$ equals the density of the unexposed film and is called 'base + fog'. The slope of the curve is referred to as the gamma ($\gamma$). Exposures in the steepest part of the curve result in the highest contrasts. The shape of the characteristic curve depends on many factors, such as film material, composition of the developer, development time, development temperature and degree of agitation during developing.

In order to obtain a high radiometric resolution, a small difference in exposure should result in a large difference in film density. Therefore the processing of the film, which was done by hand in a small reel tank, was standardized, i.e.

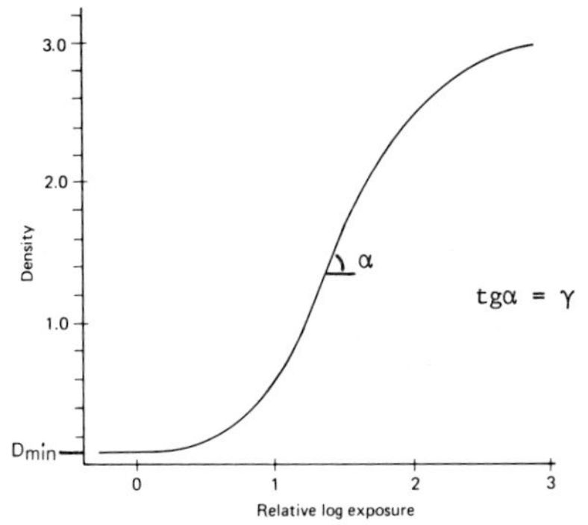


FIG. 3.9 An example of a characteristic curve.

the duration and temperature of processing, as well as agitation and drying conditions were the same. DK-50 was used as a developer. After 1 minute of prewetting, PX 2402 film was processed for 18 minutes and IR 2424 film for 16 minutes at 20 °C. During developing the solution was agitated every minute. After 2 minutes in the stop-bath, the film was put in a fixing-bath for 5 minutes and finally the film was washed for at least 8 minutes.

There are two practical ways of determining the characteristic curve:

1.) By using a sensitometric strip (step wedge). Such a step wedge is made by exposing a piece of film through a standardized wedge by means of a sensitometer. A sensitometer produces radiant energy of precisely known quantity and spectral composition. The same film material (same batch) should be used as for the material that has to be calibrated and both materials have to be processed together in order to control the whole processing. The result is a series of density steps (usually about 20) ranging from $D_{min}$ up to about a density of 3.0. The interval between the various steps is nearly constant and known (mostly 0.15 density). This implies that the relative log exposure values with such a sensitometric strip are known. The characteristic curve is obtained by measuring the densities for each step and by plotting these values against the known log exposure values.
2.) The second procedure is based on the construction of some sort of step wedge in another way (if a sensitometer is not available). Photographs are taken of a series of Lambertian reflectors with known reflectances (in this case, 5 panels with reflectances of 3%, 10%, 25%, 60% and 70%, respectively). For these reflectors a relative log exposure value can be calculated within one photograph. By changing the exposure with the relative aperture and by taking another photograph under constant irradiance conditions, new log exposure values are obtained that can be related to the former ones if the change in relative aperture is known. By taking a series of photographs of the reflectors at different relative apertures, a series of relative log exposure values is obtained. One only has to ensure that the reflectors are close to the centre of the negative, in order to avoid disturbances by light fall-off.

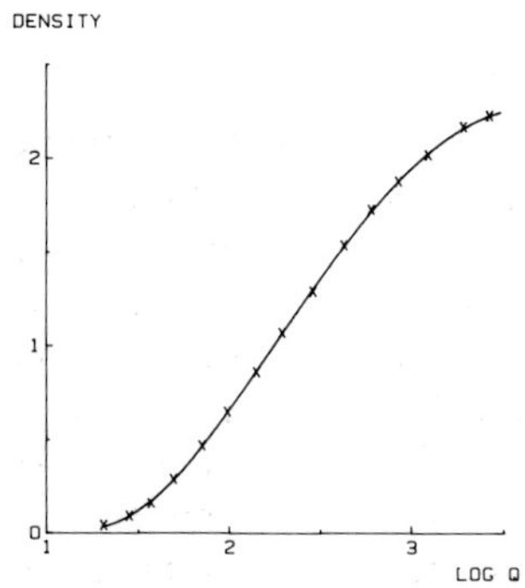


FIG. 3.10 Fourth order polynomial fit for the characteristic curve. PX 2402, batch no. 2402-282-292, W21 + W57A (green passband).

The exposure time (section 3.9) also has to be kept constant. The characteristic curve is then obtained in the same way as with procedure 1.

Both procedures were applied in this study and they yielded essentially the same results. The characteristic curve was only used as a calibration curve. Ideally, for interpolation, the characteristic curve should be described by some kind of mathematical function. It is not necessary for this latter function to be based on physical (or photographic) variables. For this reason it was decided to investigate whether a polynomial fit would be appropriate. Inherent to such a polynomial function is the fact that the fit may be bad at very low densities and at very high densities where the density of the characteristic curve is nearly constant while exposure values change. These parts of the characteristic curve were always avoided. Because of the S-shape of the characteristic curve, it was obvious that at least a third degree polynomial fit should be used for obtaining a good fit. A conventional F- or t-test can be used to compare a polynomial of degree n with one of degree n+1. If the former is not significantly worse than the latter, a polynomial of degree n-1 can be compared with one of degree n by an analogous test. This procedure was followed and a fourth degree polynomial fit was found to be optimal and was therefore used in this study. Figure 3.10 shows an example of the results of fitting a characteristic curve with a polynomial of degree four to the measurements.

The radiometric resolution of the photographic system is determined by the shape of and relative position on the characteristic curve. In this study the smallest discriminable difference in density output was 0.01. For the steepest part of the characteristic curve (density 1.0 to 1.5) this coincided with a relative radiometric resolution of 1.6% for PX 2402 and 1.2% for IR 2424. At low densities (about 0.3) or high densities (about 2.0) the curve was less steep, resulting in a relative radiometric resolution of about 2.2% for PX 2402 and 1.9% for IR 2424.

## 3.8 LIGHT FALL-OFF

Light fall-off is caused by a geometrically based decrease in illumination at the film plane with increasing distance from the centre of a photograph (called off-axis distance). If an exposure is taken of a homogeneous reflecting surface, which covers an entire film frame, the density will reach a maximum value in the centre of the frame and will decrease with increasing off-axis distance. The various factors determining this decrease in illumination have been described e.g. by Lillesand & Kiefer (1979). Figure 3.11 shows the factors causing fall-off for a film being exposed to a uniformly illuminated object. For an image point on the optical axis, exposure (E) is proportional to the area of the lens aperture

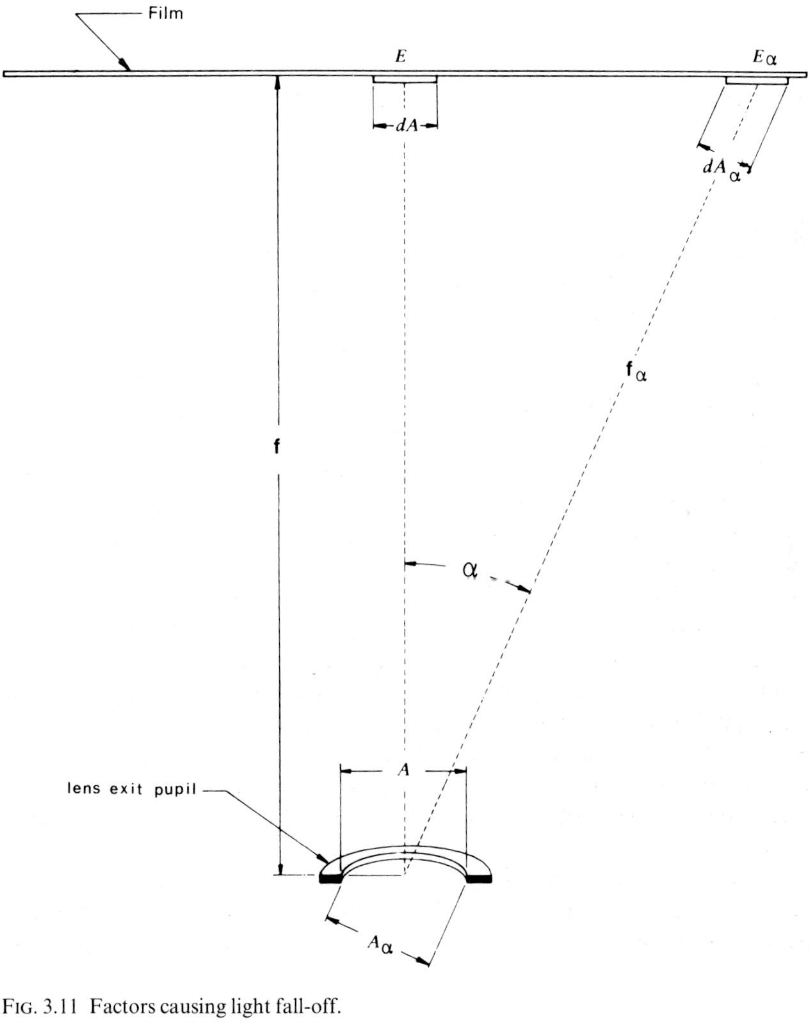


FIG. 3.11 Factors causing light fall-off.

(A) and inversely proportional to the square of the focal length of the lens ($f^2$). For an image point being exposed at an angle $\alpha$ off the optical axis, exposure ($E_\alpha$) is reduced because:

1. the effective light-collecting area of the lens (entrance pupil) ($A_\alpha$) decreases: $A_\alpha = A.\cos\alpha$.
2. the distance from the exit pupil of the lens to the focal plane ($f_\alpha$) increases: $f_\alpha = f/\cos\alpha$.
   As a result the exposure will be reduced proportionally to $\cos^2\alpha$.
3. the effective size of a film area element ($dA_\alpha$), projected perpendicular to the beam, decreases according to: $dA_\alpha = dA.\cos\alpha$.

In summary, the overall reduction in exposure will theoretically be proportional to $\cos^4\alpha$:

$$E_\alpha = E \cdot \cos^4\alpha \tag{3.2}$$

Some manufacturers of lenses try to correct for this fall-off by enlarging the entrance pupil off-axis, so the effective fall-off may be smaller than the above $\cos^4\alpha$:

$$E_\alpha = E \cdot \cos^n\alpha \tag{3.3}$$
with $n \leqslant 4$

With large relative apertures the fall-off can be augmented because of vignetting. Vignetting is the term used to describe an entrance or exit pupil rendered partly ineffective because it is masked by the lens mounts and other aperture surfaces within the lens.

A correction for light fall-off and vignetting should always be taken into account by normalizing all exposure values to the value they would have possessed if they had been at the centre of the photograph. This normalizing procedure has to be carried out for all lenses at the various aperture settings (diaphragm) used.

One approach is to take several exposures of a diffuse reflecting object (under constant illumination conditions) and, either by rotating the camera or by displacing the camera, to obtain several positions of the object along e.g. one diagonal of the frame. Because such a set-up means that variations in exposure time will induce variations in exposure, in the present study we opted to photograph an object of uniform brightness that filled a whole frame. The only concern was to obtain an object that reflected radiation diffusely.

An experiment was carried out in which a photograph was taken of an overcast sky with a diffuser (opal glass) at the focus of the lens. In another experiment a comparable configuration was used, but this time with a diffuse reflecting target, artificially illuminated, as an object. Both experiments yielded essentially the same results. Two different types of lenses were examined: a Planar 100 mm lens and a Distagon 50 mm lens. The former was used for recording under normal conditions and the latter if recording had to be carried out with a completely overcast sky and with a low cloud base. All results showed a fall-off symmetrical to the centre of the frame. Results for the various film/filter combi-

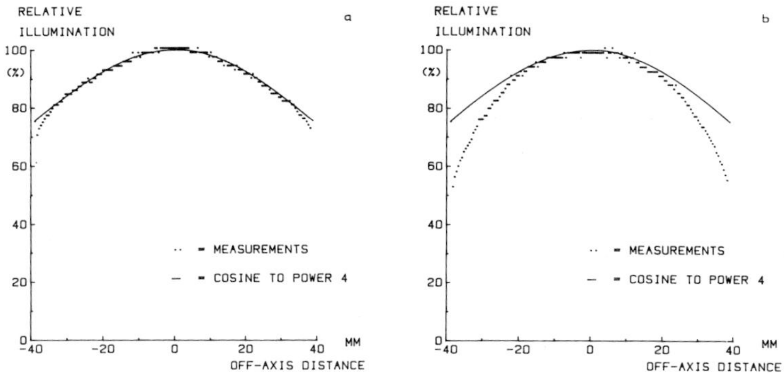


FIG. 3.12 Light fall-off for a Planar 100 mm lens. a: relative aperture 16. b: relative aperture 5.6.

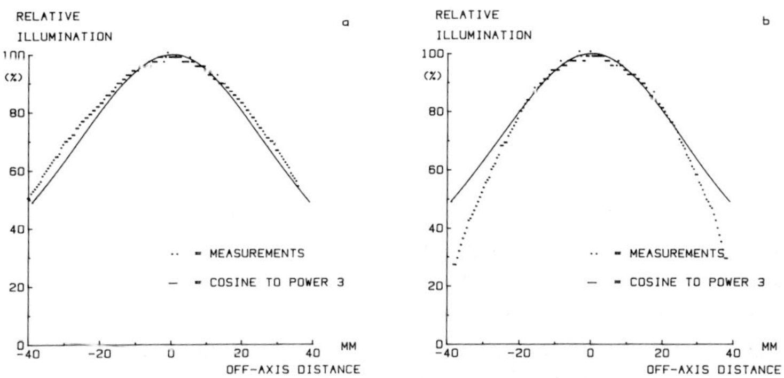


FIG. 3.13 Light fall-off for a Distagon 50 mm lens. a: relative aperture 16. b: relative aperture 5.6.

nations shown in table 3.1 did not differ. This also applied to the light fall-off results with different relative aperture settings for one lens type, except for the larger relative apertures (5.6 and larger), where vignetting occurred at the corners of the frame. This vignetting effect compounded the light fall-off effect, but the former effect did depend upon diaphragm. For the 100 mm lens, n in equation (3.3) equalled 4.0, where the angle $\alpha$ is calculated by using the focal distance (not the exit pupil distance). So the theoretical fall-off of $\cos^4\alpha$ was valid. This is illustrated in figure 3.12 for a relative aperture of 16 (a) and for a relative aperture of 5.6 (b) when there is also some vignetting. For the 50 mm lens, n in equation (3.3) equalled 3.0. Apparently, this lens had been corrected for fall-off to some extent. The fall-off for a 50 mm lens is illustrated in figure 3.13 for relative apertures of 16 (a) and 5.6 (b). In the present study only relative apertures of 5.6 or smaller were used. This implies the necessity of only correcting for light fall-off as long as at a relative aperture of 5.6 the off-axis distance is less than 25 mm (figure 3.12 and 3.13). At relative apertures smaller than 5.6 no discernable vignetting occurs.

TABLE 3.4 Values of mean exposure time (in milliseconds) and coefficient of variation (CV) of four lenses (10 measurements per setting) for several nominal exposure times.

| | nominal exposure times (milliseconds) | | | | | | | | | |
|---|---|---|---|---|---|---|---|---|---|---|
| | 2.00 | | 4.00 | | 8.00 | | 16.7 | | 33.3 | |
| | mean | CV | mean | CV | mean | CV | mean | CV | mean | CV |
| Planar No. 6180478 | 3.47 | 0.008 | 5.41 | 0.010 | 10.9 | 0.007 | 19.2 | 0.006 | 37.7 | 0.003 |
| Planar No. 6181942 | 3.54 | 0.008 | 5.66 | 0.006 | 10.1 | 0.005 | 16.5 | 0.006 | 30.1 | 0.006 |
| Distagon No. 6366552 | 4.39 | 0.004 | 5.31 | 0.005 | 10.0 | 0.005 | 17.3 | 0.005 | 30.0 | 0.007 |
| Distagon No. 6365486 | 3.11 | 0.002 | 4.85 | 0.003 | 8.5 | 0.008 | 15.3 | 0.004 | 30.4 | 0.004 |

## 3.9 EXPOSURE TIME

The lenses used have sector shutters. We tested these for consistency because if the shutter, or more precisely, the exposure time, is not consistent, this will induce varying exposure levels between successive frames. A special exposure time tester was constructed. A light-sensitive cell in the middle of the image frame of the camera triggered and stopped an electronic counter. The exposure time was displayed, the last digit being in hundreds of milliseconds. The exposure times were measured very frequently and the results are given in table 3.4. The shutters were very consistent during one day (very small coefficient of variation), and this was also true for a long time period (several months). However, the results also show that there were large differences between different shutters. The actual exposure times differed considerably from the nominal values. This had to be taken into account if the films were to be correctly exposed. During a mission one nominal exposure time only was always used.

## 3.10 RELATIVE APERTURE

The relative aperture of a lens is often indicated by N, which is defined as:

$$N = \frac{f}{d} \tag{3.4}$$

f = focal length of the lens
d = diameter of the entrance pupil.

The irradiance at an axial point in the image plane of a camera is (derived from Slater, 1980):

$$E_I = \frac{\pi \cdot L_L}{4 \cdot N^2} \tag{3.5}$$

$E_I$ = image irradiance
$L_L$ = radiance incident over the exit pupil in a perfect system without reflectance, absorption and scattering losses.

The image irradiance $E_I$ is inversily proportional to the square of the relative aperture. In photography, different levels in exposure are normally established in sequential multiples of two. Thus N varies as $(\sqrt{2})^n$ with n = 1,2,... (1.4, 2, 2.8, 4, 5.6, 8, 11, 16, 22) and log exposure varies as multiples of log 2 = 0.30.

The relative aperture of a lens at a certain setting will be a constant value. However, it is possible that after changing this setting into another value and then going back to the original setting this new value will differ from the original one. During one mission all objects were photographed with each film/filter combination successively, without changing the settings for one combination. An experiment was carried out to ascertain whether deviations could occur. It was also investigated whether a nominal full stop in relative aperture corresponds with the measured difference in exposure.

In the experiment, exposures were taken of a target containing five diffuse reflectors of different levels of greyness (cf. section 3.7). At one constant setting for the exposure time the relative aperture was varied in the order: 4–5.6–8–16–22 and back again to 4 (for distinguishing a possible difference between e.g. 4 → 5.6 and 8 → 5.6). This was repeated for all 4 lenses mentioned in table 3.4. The characteristic curve was ascertained using a sensitometric strip. Densities were measured and, subsequently, corresponding log exposure values were ascertained at the various relative apertures.

An important result was that the log exposure or exposure value at a certain relative aperture at a certain lens was very constant (taking differences resulting from variations in exposure time and measurement inaccuracies into account). It was immaterial whether a certain relative aperture value was approached from a large or small setting. Secondly, the difference in exposure between relative apertures 8, 11, 16 and 22 matched precisely (within 0.01 Δlog exposure) multiples of two each time (Δlog exposure = 0.30). However, the difference between 5.6 and 8 was slightly less than a factor of two for all lenses (Δlog exposure was about 0.28, which corresponds to a factor 1.9). This discrepancy from a factor 2.0 was still small enough to allow it to be ignored in establishing correct exposure levels during missions. For all lenses the difference in exposure between relative apertures 4 and 5.6 was much less than a factor of two. However, the relative aperture 4 was never used during missions, because of the large influence of vignetting at that (full) aperture.

## 3.11 Focal setting

The focal length of an ordinary lens for infrared is greater than the focal length for visible radiation (Manual of photography, 1978). An experiment was carried out in order to ascertain which focus setting should be used with different lenses for the various film/filter combinations. This was done by photographing various objects at large distance (infinity) with several distance settings of the lens.

It was found that for both the Planar 100 mm lens and the Distagon 50 mm lens the focus setting should be, as expected, at infinity when exposing films in the visible region of the spectrum. In the near infrared region (700–900 nm) the Planar 100 mm lens should be set at 20 metres and the Distagon 50 mm lens should be set at 7.5 metres.

## 3.12 Radiometric calibration and reference targets

In chapter 4 the process of reflectance and the influence of the atmosphere are described mathematically up to the stage of radiance outside the sensor. Photography registers the radiant energy reaching the film plane. Equation (3.5) described in simple form the image irradiance in a system without losses, but a more specific form is given in the following equation (after: Slater, 1980), in which transmittance of the optical system, light fall-off, vignetting and camera flare are also taken into account:

$$E_I = \frac{K_N(\alpha) \cdot \cos^n\alpha \cdot \pi \cdot L_s \cdot \tau_o}{4 \cdot N^2} + E_f \qquad (3.6)$$

$K_N(\alpha)$ = vignetting factor
$\cos^n\alpha$ = light fall-off factor
$L_s$ = radiance measured outside the sensor
$\tau_o$ = transmittance of the optical system
$E_f$ = camera flare.

If one uses energy values the exposure time (t) also has to be included, and equation (3.6) changes into:

$$Q_I = \frac{K_N(\alpha) \cdot \cos^n\alpha \cdot \pi \cdot L_s \cdot \tau_o \cdot t}{4 \cdot N^2} + Q_f \qquad (3.7)$$

$Q_I$ = radiant energy detected at the image plane
$Q_f$ = radiant energy camera flare.

Combining equation (3.7) with equations (4.1) and (4.3) (chapter 4) for a specified object (o) results in (after: Slater, 1980):

$$Q_I = \frac{K_N(\alpha) \cdot \cos^n\alpha \cdot \tau_o \cdot t}{4 \cdot N^2} \cdot (E_o \cdot r_o \cdot \tau_a + E_a) + Q_f \tag{3.8}$$

$E_0$ = irradiance incident on the object
$r_0$ = reflectance of the object
$\tau_a$ = atmospheric transmittance
$E_a$ = irradiance from the atmosphere.

When avoiding vignetting effects and after correction for light fall-off, equation (3.8) reduces to:

$$Q_I = \frac{\tau_o \cdot t}{4 \cdot N^2} \cdot (E_o \cdot r_o \cdot \tau_a + E_a) + Q_f \tag{3.9}$$

Camera flare conditions are constant for identical camera systems and identical distributions of reflectance elements in the field of view. So, for a given mission with constant atmospheric conditions and camera setting for a specified passband, all terms in equation (3.9) can be regarded as constants except $Q_I$ and $r_0$, which depend on the object. This implies that equation (3.9) can be rewritten as:

$$Q_I = a + b \cdot r_o \tag{3.10}$$

a and b being constants depending on passband.

In order to ascertain a and b for a specified passband, two types of reference targets were set up in the field during missions: seven of coloured curtain cloth (3.6 m × 3.6 m) and three grey Lambertian targets (5.0 m × 5.0 m) of aluminium, each formed by eight plates coated with paint, especially selected by the TNO Paint Research Institute (Jonkers, 1982). The reflectance characteristics of the targets as measured with the field spectroradiometer (cf. section 2.4) are given in figure 3.14. Only targets with a uniform reflectance in a relevant passband were used in this study for calibration, because the boundaries of the passbands were not very sharp. Moreover, reflectance values that were out of range (much lower or higher than those of soil or vegetation) were not used. The reflectance values in the passbands used with aerial photography for the targets used for calibration are given in table 3.5.

During each mission the targets were recorded with the various film/filter combinations at the same camera setting as the other objects (and under the same atmospheric conditions). They were measured by using the same densitometric procedure as described previously. The inverse of equation (3.10) was then used for obtaining calibrated reflectances for each object:

$$r_0 = A + B \cdot Q_I \tag{3.11}$$

with constants A and B.

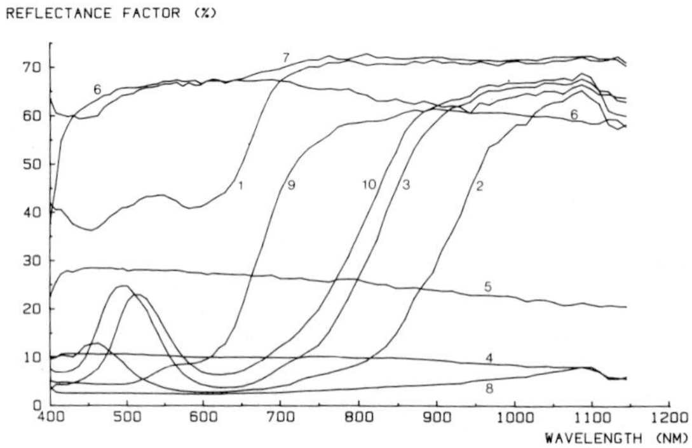


FIG. 3.14 Reflectance characteristics of ten targets as measured with the field spectroradiometer on 27 May 1982.

| | |
|---|---|
| 1 = grey target | 6 = 60% target |
| 2 = blue target | 7 = white target |
| 3 = green/blue target | 8 = black target |
| 4 = 10% target | 9 = red target |
| 5 = 25% target | 10 = green target |

Targets 4, 5 and 6 are made of aluminium, the remaining ones of cloth.

TABLE 3.5 Reflectance values (%) of the targets used for calibrating the multispectral photography recordings (in the specified passbands).

| target | PX 2402<br>W21 + W57A<br>green | PX 2402<br>W70<br>red | IR 2424<br>W87C<br>infrared |
|---|---|---|---|
| 1 grey | – | – | 71.0 |
| 2 blue | 3.03 | 3.55 | – |
| 3 green/blue | – | – | – |
| 4 10% | 10.2 | 9.9 | 9.6 |
| 5 25% | 27.9 | 26.9 | 24.3 |
| 6 60% | – | – | – |
| 7 white | – | – | 72.0 |
| 8 black | 2.32 | 2.63 | – |
| 9 red | 8.47 | – | 60.9 |
| 10 green | – | – | – |

An example of the results of this calibration is given in figure 3.15. It is noteworthy that the constant A always is negative, since $Q_l$ will be positive at reflectance zero because of irradiance from the atmosphere.

## 3.13 CHECKING REFLECTANCE FACTORS OBTAINED WITH MSP

During the first part of the 1982 growing season, reflectance measurements were carried out at the ir. A.P. Minderhoudhoeve in field trial 116 (cf. chapter

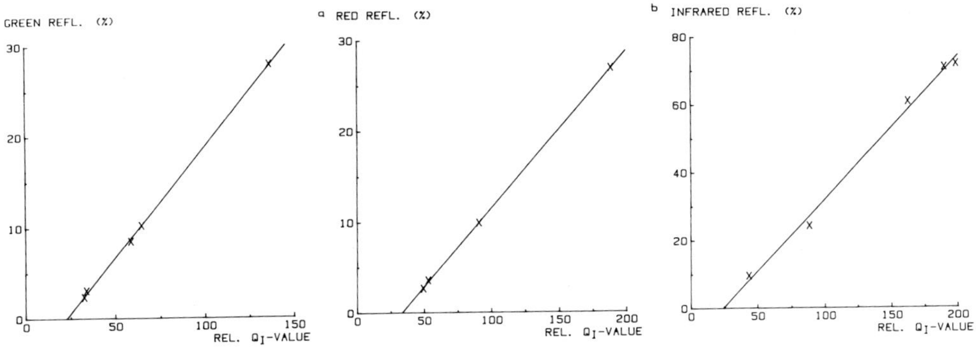


FIG. 3.15 Camera calibration and atmospheric correction on 27 May 1982.
a: green passband : $r_0 = -\ 5.56 + 0.245\ Q_I$
b: red passband : $r_0 = -\ 5.77 + 0.173\ Q_I$
c: infrared passband : $r_0 = -10.25 + 0.424\ Q_I$

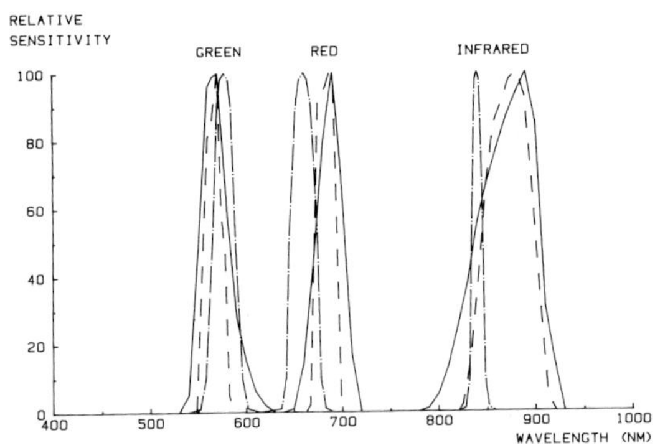


FIG. 3.16 Comparison of the spectral passbands of the MSP system (—), field spectroradiometer (---) and hand-held radiometer (.-).

6), using the field spectroradiometer (FSM) (section 2.4). In order to simulate the wider passbands used with multispectral aerial photography (figure 3.3) a weighted average of some narrow passbands of the field spectroradiometer was calculated for the green, red and infrared (figure 3.16). During the second part of the season such measurements were taken using the hand-held radiometer (HRM) (section 2.4). These measurements were compared with the reflectance factors of the same plots, which had been ascertained by means of multispectral photography (MSP). The passbands used with all three systems are illustrated in figure 3.16. Results of the comparison for the green, red and infrared passband are illustrated in figures 3.17 and 3.18 for the crops sown early and late, respectively. Average reflectance factors are shown for each passband for the early- and late-sown crops. These results show that the reflectance factors obtained

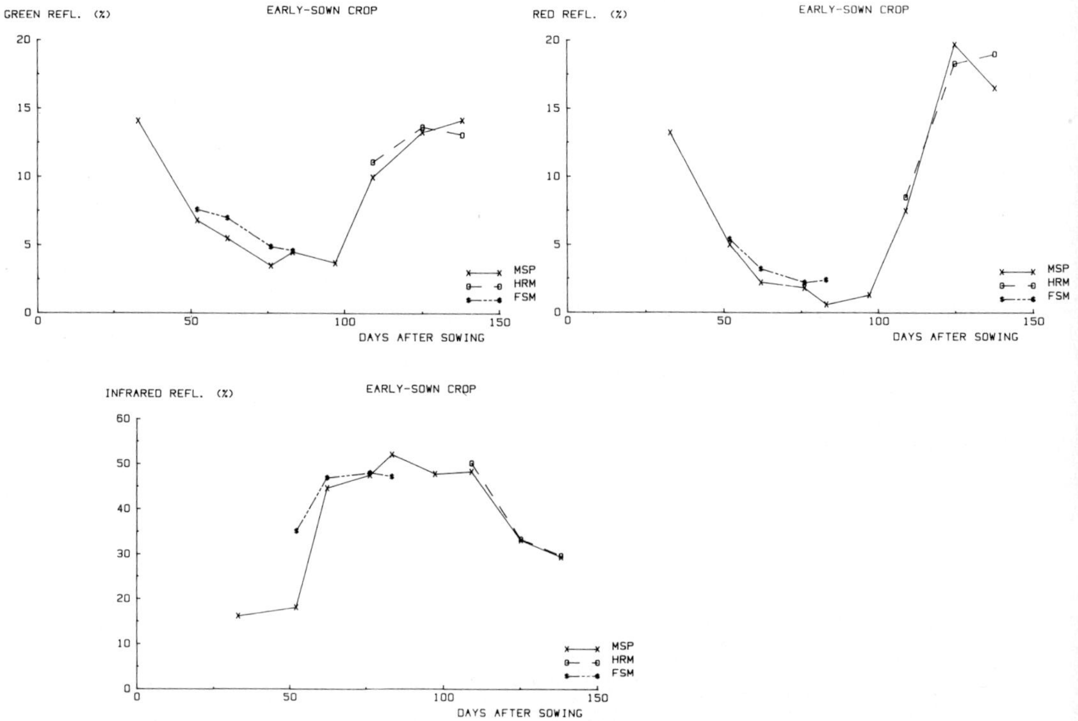


FIG. 3.17 Comparison of reflectance factors ascertained by means of multispectral photography (MSP) with those obtained by means of a hand-held radiometer (HRM) and a field spectroradiometer (FSM) in field trial 116 in 1982 for the early-sown crop.

by multispectral aerial photography coincide with those obtained by means of radiometers in the field.

Results obtained during the 1983 growing season in field trials 116 and 92 are illustrated in appendices 3 and 4, respectively.

On 13 July 1982 a multispectral scanning (MSS) mission was carried out at the ir. A.P. Minderhoudhoeve. In field trial 116, reflectance factors per plot were calculated for channels 5, 7 and 9. The reference targets were used for converting radiances into reflectance factors (cf. section 4.2). On the same date an MSP mission was also carried out and reflectance factors per plot were calculated for the green, red and infrared passbands.

Results of the comparison of MSP with MSS are shown in figure 3.19 for the 36 plots in field trial 116 in 1982. Noteworthy is the different level of red reflectance with MSS compared to that with MSP (results obtained by MSP coincided with those obtained by means of the hand-held radiometer). With MSS the reflectance in channel 7 was higher than that in channel 5, which is contrary to the results with MSP. An explanation for this phenomenon may be that the position of channel 7 in the spectrum has shifted towards larger wavelengths, incorporating part of the reflectance of vegetation in the infrared.

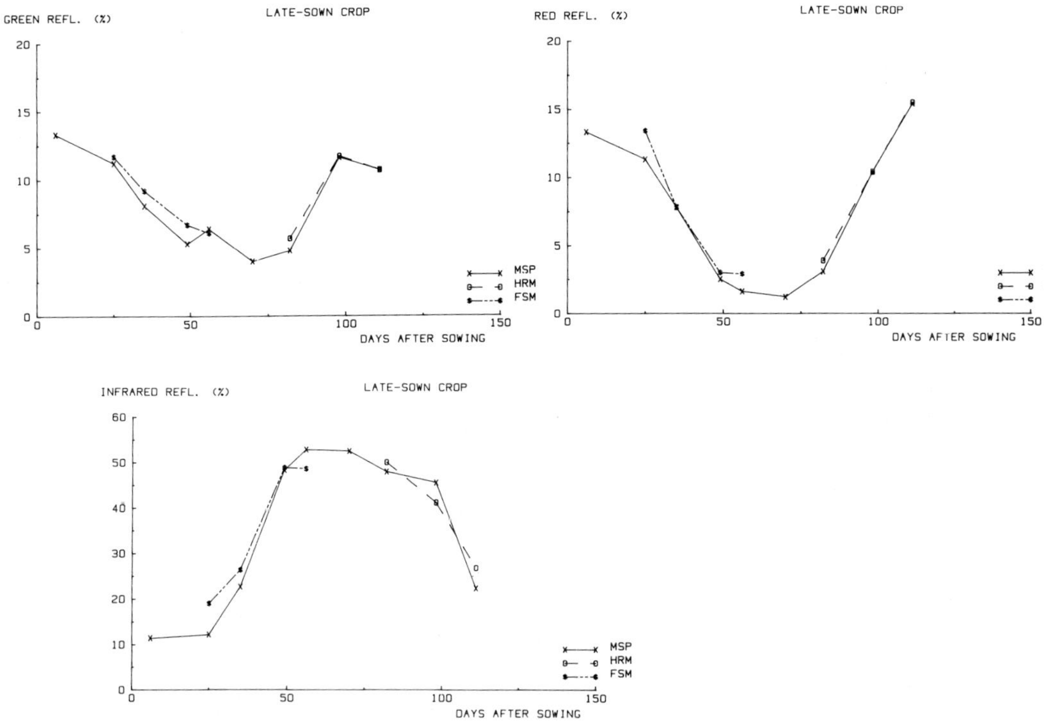


FIG. 3.18 Comparison of reflectance factors ascertained by means of multispectral photography (MSP) with those obtained by means of a hand-held radiometer (HRM) and a field spectroradiometer (FSM) in field trial 116 in 1982 for the late-sown crop.

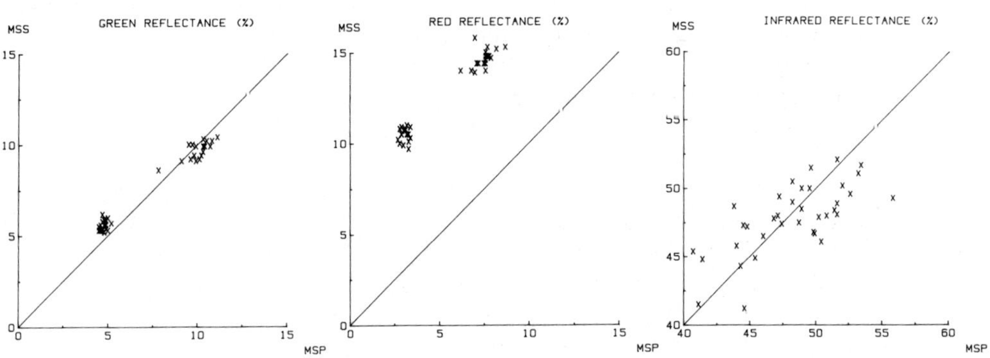


FIG. 3.19 Comparison of multispectral photography (MSP) with multispectral scanning (MSS) in field trial 116 on 13 July 1982.

Since the infrared reflectance was high, the integrated reflectance of the red and part of the infrared (e.g. 650–730 nm instead of 650–700 nm) will be relatively high.

It can be concluded that the procedure of MSP described in this chapter yields calibrated reflectance factors with high spectral resolution. The reflectance factors coincide with those obtained with MSS (except in the red) and with those obtained in the field by means of a hand-held radiometer and a field spectroradiometer.

## 3.14 Summary

Black and white multispectral aerial photography was carried out by taking vertical photographs from a single-engine aircraft, equipped with Hasselblad cameras with Planar 100 mm lenses (or in some cases Distagon 50 mm lenses), at intervals of about two weeks. A Minolta photometer was used for correcting the camera settings for differences in illumination between missions, and the expected reflectance level of the objects was also taken into account for obtaining correct levels of exposure. During each mission reference targets were set up in the field and these were photographed together with the field trials.

Missions were only carried out under nearly constant atmospheric conditions, and during one mission camera settings (exposure time and relative aperture) were not altered for a certain film/filter combination. Film (black and white) and filters were selected in order to acquire exposures in green, red and infrared passbands. The development of the films was standardized.

Densities were measured by means of an automatized Macbeth TD-504 densitometer. After converting the densities to radiant energy values by means of the characteristic curve, compensation for off-axis imagery was made for each image point (only for light fall-off; vignetting was avoided). There is a linear relationship between this corrected radiant energy value and the reflectance of the object. The parameters were ascertained by means of calibrated reference targets; subsequently, calibrated reflectances were calculated for each object.

Reflectance values ascertained in this way with multispectral aerial photography appear to be well calibrated when compared with MSS and radiometers in the field, offering information that is only influenced by the object. This allows multitemporal comparison and analysis to be done.

# 4 REMOTE SENSING OF VEGETATION

## 4.1 Introduction

As the preceding chapters have shown, black and white multispectral aerial photography appears to be an appropriate remote sensing system for use in agricultural field trials. Such a system ascertains reflectances in the optical region of the electromagnetic spectrum. In this chapter first of all spectral reflectance will be defined and its relation to the signal detected by the sensor at an aerial platform will be given. Then the main crop characteristics that can be measured will be described and the influence of the soil will be considered. Finally, several indices (which are functions of the reflectances in the various passbands) and models that may be used for estimating crop characteristics will be reviewed.

## 4.2 Spectral reflectance

As stated in section 2.2, electromagnetic radiant power reaching the earth's surface will result in three interactions: transmission, absorption and reflection. In remote sensing the latter is measured. However, the three of them are interrelated because of the principle of conservation of energy. The amounts of radiant power involved in these interactions will be expressed as portions of the total incident radiant power. For the reflected radiant power this portion is called the spectral reflectance. Spectral hemispherical reflectance at wavelength increment $(\lambda, \lambda + d\lambda)$ is defined as the ratio of reflected to incoming radiant power of wavelength increment $(\lambda, \lambda + d\lambda)$ per unit of surface area (after: Bunnik & Verhoef, 1974). Transmittance and absorptance are defined analogously. The spectral reflectance is essentially a function of wavelength and of the properties of the object. This means that spectral reflectance of different objects may be almost equal at certain wavelengths, but may be clearly different at other wavelengths. This is important for discriminating between objects.

An object may reflect radiation uniformly in all directions (so-called ideal diffuse or Lambertian reflector). Natural surfaces will not be Lambertian reflectors, but they approximate this condition. For such a surface the reflected radiance in the direction of the sensor is defined by the radiant power reflected per unit of the reflecting area in the direction of observation, per unit of solid angle and per unit of wavelength increment, and is given by (after: Bunnik, 1978):

$$L_o = r_o \cdot \frac{E_o}{\pi} \qquad (4.1)$$

$L_0$ = radiance reflected by the object in the direction of the sensor in $Wm^{-2}sr^{-1}\mu m^{-1}$

$r_0$ = directional reflectance factor of the object
$E_0$ = hemispherical irradiance in $Wm^{-2}\mu m^{-1}$.

The constant $\pi$ appears because it is the directional reflectance factor of an object instead of the hemispherical reflectance that is ascertained by a remote sensing system. This directional reflectance factor is defined as the radiance of the object in a certain direction under certain irradiation conditions, relative to the radiance of an ideal, white, diffusing surface, in the same direction and under the same irradiation conditions (after: Verhoef & Bunnik, 1975). Instead of 'reflectance factor' the term 'reflectance' is often used, although the former is the correct term. Equation (4.1) can be rewritten as:

$$r_o = \frac{\pi \cdot L_o}{E_o} \tag{4.2}$$

Often, the surface radiance of an object and the radiance detected by the sensor at a remote distance are unequal, because of the interference of the atmosphere. Firstly, the atmosphere attenuates the radiance reflected by an object and, secondly, it adds a certain amount of radiance because of directional scattering in the atmosphere (figure 4.1). This latter term is often called path radiance (after: Slater, 1980).

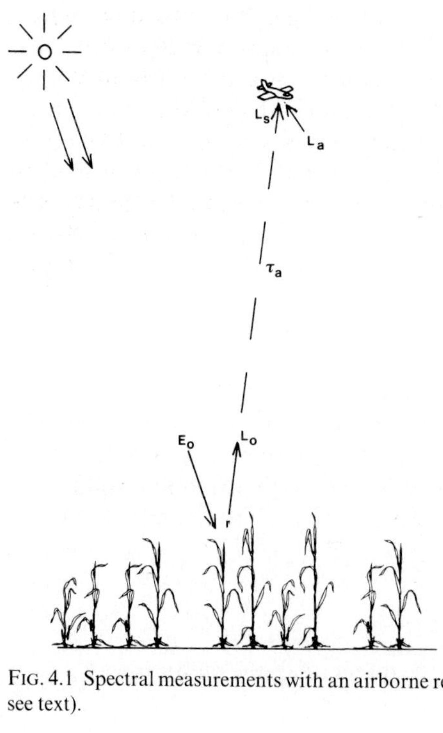


FIG. 4.1 Spectral measurements with an airborne remote sensing system (for explanation of symbols see text).

$$L_s = L_0 \cdot \tau_a + L_a \tag{4.3}$$

$L_s$ = radiance detected by the sensor
$\tau_a$ = atmospheric transmittance
$L_a$ = path radiance (atmosphere).

Combining equations (4.2) and (4.3) results in:

$$r_o = \frac{\pi \cdot (L_s - L_a)}{\tau_a \cdot E_o} \tag{4.4}$$

The radiance reaching the sensor then enters the sensor. The radiance may then be transformed in some way before it is finally detected. In section 3.12 this procedure was described for an aerial photography system.

All the terms in equation (4.1) to (4.4) will depend on wavelength.

## 4.3 Main crop characteristics measurable with remote sensing

Figure 4.2 shows the spectral reflectance of some distinct objects. In the visible region, vegetation absorbs much radiation and shows a relatively low reflectance. This is especially true in the red region, because of the large absorption of this radiation by the chlorophyll in the leaves. In the near infrared region the opposite occurs. The spectral reflectance of leaves in this region is high. The reflectance and transmittance of a green leaf are approximately equal (e.g. Goudriaan, 1977; Youkhana, 1983). This is shown in figure 4.3.

Almost all soils show a continuous increase in reflectance from 400 nm up to 1300 nm without sudden ups or downs (figure 4.2). Two dips appear at approximately 1450 nm and 1950 nm because water present in the soil absorbs incoming radiation. There appears to be a clear contrast in reflectance between bare soil and vegetation in the visible (red and green) region and also in the infrared region.

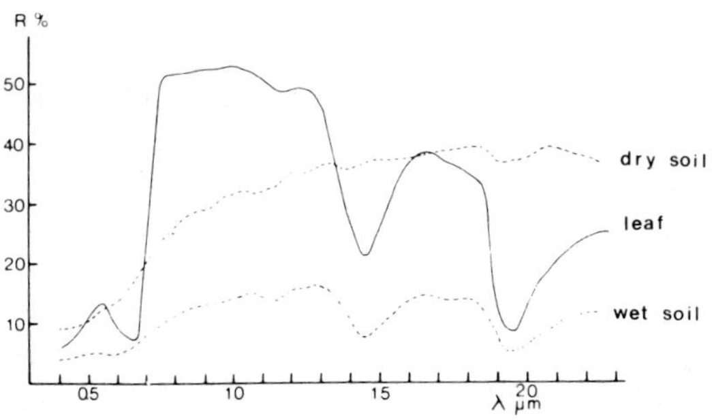


FIG. 4.2 Spectral reflectance of a green wheat leaf compared with the reflectance of dry and moist sandy loam soils (from: Bunnik, 1978).

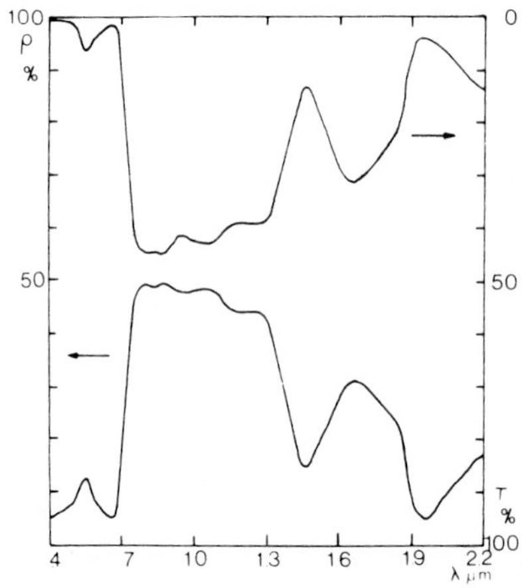

FIG. 4.3 Reflectance (ρ) and transmittance (T) of a single green leaf, used in canopy reflectance simulations by Bunnik (1978).

In the visible region of the electromagnetic spectrum, reflectance and transmittance of a green leaf are in the order of 10 % or less. This means that absorptance is at least 80 %. If 10 % of the incoming radiation is reflected by the upper leaf, the contribution to the measured reflectance of a second leaf underneath this upper leaf would be only 1 % of the reflectance of the upper leaf (figure 4.4). For a crop canopy this implies that in the visible region only the reflectance of the upper layer of leaves determines the measured reflectance of that canopy. So, the large contrast in reflectance between bare soil and vegetation in the visible region will be most suitable for estimating soil cover, because this reflectance will attain a minimum value (maximum absorption) at the moment of complete soil cover.

Figure 4.5 shows an example of model simulations by Bunnik (1978) of the spectral reflectance for vegetation, including soil background, with different LAI. In the visible region chlorophyll absorption causes a drop in reflectance as LAI increases. For high LAI values this drop is minor. In the infrared region a noticeable increase in reflectance occurs as LAI increases (also at the larger LAI values).

In the near infrared region of the electromagnetic spectrum, reflectance and transmittance of a green leaf are each about 50 %, and there is hardly any infrared absorptance by a green leaf. In this situation leaves or canopy layers underneath the upper layer (or soil underneath it) contribute significantly to the total measured reflectance (figure 4.4). In the rather simple situation of the reflectance and the transmittance both being 50 %, the contribution from a second leaf (or layer) is about 15 % of the incoming radiation, a value that is noticeable. The contribution from subsequent leaves (or layers) decreases exponentially. This multiple reflectance suggests that the infrared reflectance may be a suitable estimator of LAI.

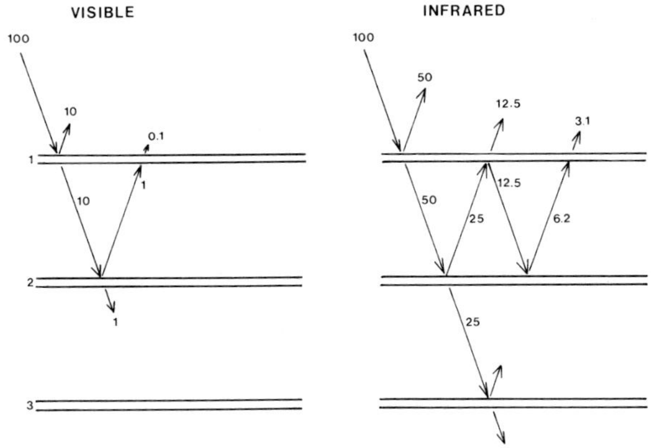


FIG. 4.4 Schematic representation of the reflectance and transmittance of radiation by canopy layers in the visible and infrared regions, respectively.

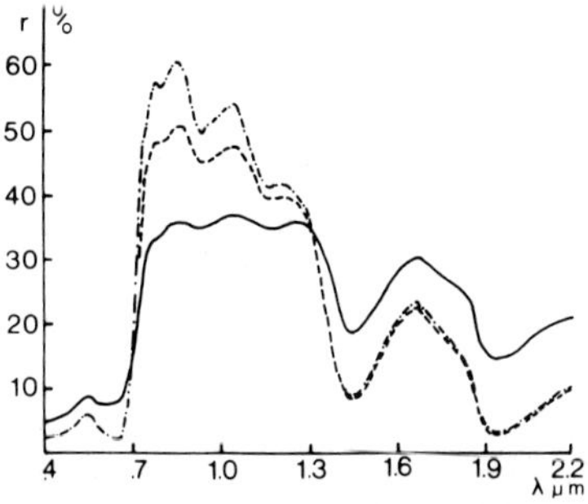


FIG. 4.5 Spectral reflectance of vegetation with LAIs of 0.5 (—), 2.0 (---) and 8.0(-·-) on a dry sandy loam soil (from: Bunnik, 1978).

The above representations (figure 4.4) are an oversimplification of reality, but they explain the general concepts.

In summary, it can be concluded that soil cover may best be estimated from reflectance in the visible region and that LAI may best be estimated from infrared reflectance.

## 4.4 Estimation of other crop characteristics

### *4.4.1 Leaf colour*

Colour refers to an aspect of visual experience. It describes a form of perception. Unfortunately, it is very difficult to obtain precise, quantitative measurements of the magnitude of perceptions. On the other hand, it is less difficult to obtain precise measurements of spectral characteristics of objects. The colour of an observed object is determined by the spectral power distribution of the radiant source and the reflectance of the object. In this study the colour in terms of the reflectance in distinct wavelength bands was of interest, since comparisons per wavelength band were primarily made. So, leaf 'colour' refers to leaf reflectance in a few passbands.

Leaf colour is important as an indicator of different phenomena. It may, for instance, indicate a deficiency or a surplus of certain nutrients; it may also indicate differences in plant structure or a certain disease or senescence of a crop. Leaf colour may be measured as the reflectance in a green and red passband (Bunnik, 1978).

At the beginning of the season, when soil cover is incomplete, it is difficult to estimate leaf colour explicitly from remote sensing measurements unless one knows the reflectance of the soil. At that stage, differences in leaf colour constitute only a minor contribution to the measured reflectance in a visible passband. Because leaf reflectance in a visible passband is at a low absolute level compared to bare soil, possible differences in leaf colour are dominated by soil background.

At complete soil cover the measured reflectance is at a low absolute level and small absolute differences in leaf reflectance in visible passbands should be detectable. The cause of possible differences in leaf colour is usually not evident from spectral reflectance values and is left to the interpreter.

At the end of the growing season, large increases in leaf reflectance in visible passbands will occur because of natural leaf senescence. However, leaf senescence may also be caused by disease.

### *4.4.2 Yield*

Often one is interested in predicting yield. This yield may be expressed as e.g. grain yield, potato yield or sugar yield, depending on the plant part one is going to harvest. Also, estimating dry matter weight at different periods during the growing season may be important.

During a large part of the growing season the LAI is clearly related to dry matter weight. During the first part of the growing season LAI and dry matter weight will both increase and correlate strongly. After maximum LAI has been reached, LAI will decrease because of senescence of leaves. However, total dry matter weight will not decrease at all, or at most only to a small extent.

Reflectance, especially infrared reflectance, is primarily related to LAI. Because LAI correlates strongly with dry matter weight during part of the growing season, reflectance measurements can also be useful for estimating dry matter weight. The relationship, however, is only indirect. At the end of the season the relationship between dry matter weight and reflectance values will disappear, so that it will be difficult to estimate dry matter weight from reflectances at that moment. Only during that part of the season in which LAI and dry matter weight correlate clearly can dry matter weight be estimated via the relationship between reflectance and LAI.

Yield at the end of the season, i.e. yield of the plant part (or organ) one is interested in, is always the result of photosynthetic activity of the green leaves. So not only the absolute value of LAI at a certain moment (or the maximum value of LAI) is important, but also the time period in which a certain LAI value exists. In agriculture the integrated value of the green LAI over time is often used to estimate the final yield. Hence there will only be an indirect relationship between reflectances and final yield.

In general it may be stated that reflectance values should be used primarily for estimating soil cover and LAI. They should not be used for directly estimating yield or dry matter weight. However, reflectance values may be used indirectly to estimate yield or dry matter weight if one knows the relationship of LAI with yield or dry matter weight.

*4.4.3 Diseases*

It is not always possible to detect diseases or other stresses in crops directly from reflectance values. Whether a disease (or stress) will be detected in an early stage depends on how it occurs in the vegetation. The following four situations may occur when disease is present in a crop:

1. A disease may be very difficult to detect at an early stage, e.g. fungi that attack the roots of plants or the base of their stems, but do not at first cause any reaction in the plant itself. At some point the plant may collapse suddenly, and this may then be detected by reflectance measurements–although it may be too late to save the crop.
2. A disease may cause a slight difference in leaf colour, but no immediate slowing down of growth. This may be detected by means of reflectance measurements in visible passbands.
3. A disease may cause a slight slowing down of growth, but no apparent change in leaf colour. This may be detected by means of reflectance measurements in an infrared passband because the value of the LAI will be influenced.
4. A combination of 2. and 3. will often occur. In this situation it will be possible to detect the effect of a disease by measuring in a visible passband and in an infrared passband.

*4.4.4 Lodging*

Lodging may occur when the stems of cereals are too long or too weak to bear the heavy ear. They will bend and finally the stem may break. This can happen as a result of an overdose of nitrogen. Disease attack of the roots or the stem may also be a cause.

Lodging may cause the whole canopy structure to change. More stems and fewer leaves will become visible from above. This will affect the reflectance in a visible passband. At first, lodging will not influence the LAI, but because the leaves are stacked close to one another near the soil surface, the photosynthetic activity will decrease considerably. This will induce a decreased yield at the end of the season. Because there will only be a minor decrease in LAI, there will be little change in infrared reflectance.

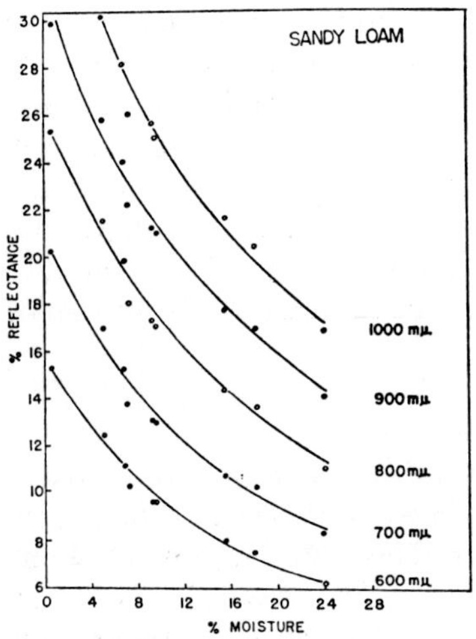


FIG. 4.6 Reflectance as a function of soil moisture content for various wavelengths (from: Bowers & Hanks, 1965).

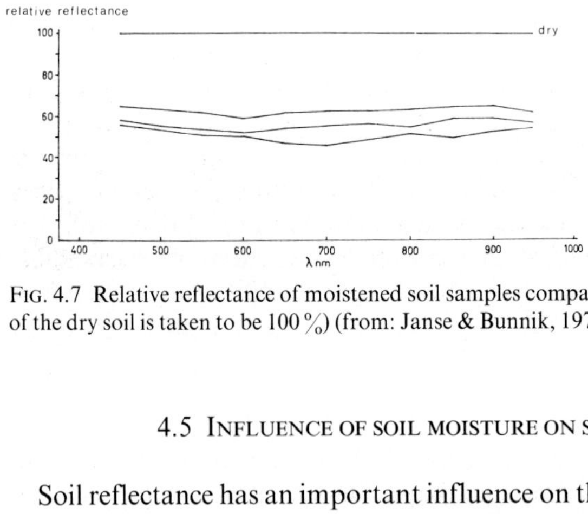


FIG. 4.7 Relative reflectance of moistened soil samples compared with a dry sample (the reflectance of the dry soil is taken to be 100 %) (from: Janse & Bunnik, 1974).

## 4.5 INFLUENCE OF SOIL MOISTURE ON SOIL REFLECTANCE

Soil reflectance has an important influence on the relationship between reflectance and soil cover and on the relationship between reflectance and LAI. At low soil cover, soil reflectance contributes strongly to the measured reflectance in the different passbands. For a given soil type, soil moisture will be the main factor determining soil reflectance. It is important to know in which way soil moisture influences soil reflectance and whether this influence depends on wavelength in the range of the electromagnetic spectrum under consideration.

An increasing moisture content of the soil causes the reflectance to decrease (Bowers & Hanks, 1965). However, the relative effect of soil moisture on the reflectance at distinct wavelengths is similar (figure 4.6). Janse & Bunnik (1974) noticed that reflectance decreases with increasing soil moisture content, but this decrease is almost independent of wavelengths between 400 nm and 1000 nm for a sandy soil (figure 4.7). This means that the ratio of the reflectance in two passbands in this interval is nearly independent of soil moisture content. Condit (1970) classified all soil types into three main classes on the basis of their reflectance pattern. The ratio of red and green reflectance, for instance, ranged from 1.5 to 1.7 in class I, from 1.1 to 1.4 in class II and from 2.0 to 3.0 in class III. For a specific soil type this ratio is independent of soil moisture content. Results obtained by Stoner et al. (1980) confirm that the ratio of the reflectance in two passbands is independent of the soil moisture content.

## 4.6 Indices for estimating crop characteristics

The green, red and infrared reflectances may be used as variables for estimating crop characteristics such as soil cover, LAI or yield. The great variation found in several studies on the relationship between reflectance and crop characteristics is not only caused by differences in crop characteristics, but also by differences in soil background or atmospheric conditions. This complication is particularly prevalent when spatial and temporal analysis of reflectances is performed. Much research has been aimed at establishing combinations of the reflectances in different wavelength bands, to minimize these undesirable disturbances. Most of these investigations relate to Landsat bands 4 (MSS4: 500-600 nm), 5 (MSS5: 600-700 nm), 6 (MSS6: 700-800 nm) and 7 (MSS7: 800-1100 nm). However, when using some combination of reflectances, one should be careful not to lose sensitivity to variations in LAI or dry matter weight after complete soil cover has been reached. This also means that the infrared reflectance should play a dominant role in such a combination.

The earliest investigations involved the infrared/red ratio (e.g. Rouse et al., 1973, 1974):

$$\frac{MSS7}{MSS5} \qquad (4.5)$$

Rouse and his colleagues found that this ratio was especially useful for estimating crop characteristics by correcting the radiances measured by earth-observation satellites (e.g. Landsat), for eliminating seasonal sun angle differences and for minimizing the effect of atmospheric attenuation. The same authors also used the 'vegetation index' (VI) for this purpose. This index is defined as:

$$VI = \frac{MSS7 - MSS5}{MSS7 + MSS5} \qquad (4.6)$$

In order to avoid negative values a transformed vegetation index (TVI) was also used:

$$TVI = \sqrt{VI + 0.5} \qquad (4.7)$$

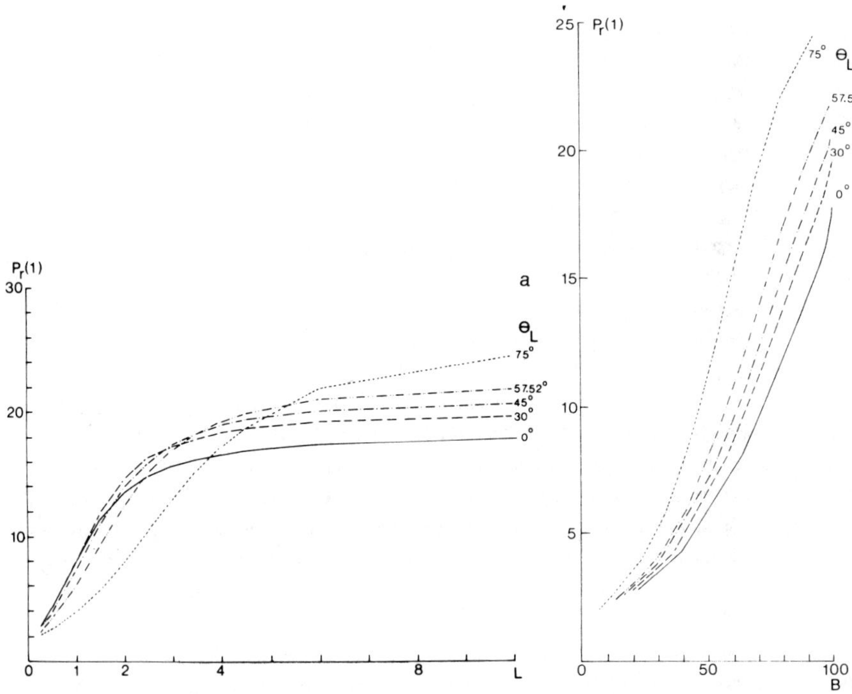


FIG. 4.8 Ratio of infrared and red reflectance ($P_r(1)$) as a function of (a) LAI (L) and (b) soil cover (B) for a wheat crop with different leaf angles ($\theta_L$) on a dry sandy loam soil (from: Bunnik, 1978).

These indices have also been used by many other investigators, e.g. Ahlrichs & Bauer (1983), Asrar et al. (1984), Curran (1980, 1981, 1982a, 1982b), Hatfield (1981), Hatfield et al. (1984), Holben et al. (1980), Pinter et al. (1981), Tucker (1979, 1980).

The value of the infrared/red ratio at complete soil cover (large LAI) is highly dependent on the contribution of the red reflectance, because this value will be at a small absolute level. However, the discussions in section 4.3 indicate that from complete soil cover onwards red reflectance will hardly be influenced by a further increase in LAI. So, it is very undesirable for red reflectance at that stage to have a large influence on any index used for estimating LAI or dry matter weight. E.g. in mid-June 1982 both the LAI and the infrared reflectance for field trial 116 were hardly influenced by the sowing date. However, in that year there was a significant difference in red reflectance for both sowing dates (about 1 % as opposed to 2 %). This resulted in a large difference in infrared/red ratio.

Model simulations done by Bunnik (1978, 1981) confirm that these indices may be useful for estimating soil cover, but are only slightly sensitive for variations in LAI or dry matter weight after complete soil cover has been reached (figures 4.8 and 4.9). This is also confirmed by the results of e.g. Asrar et al. (1984), Hatfield et al. (1984), Holben et al. (1980).

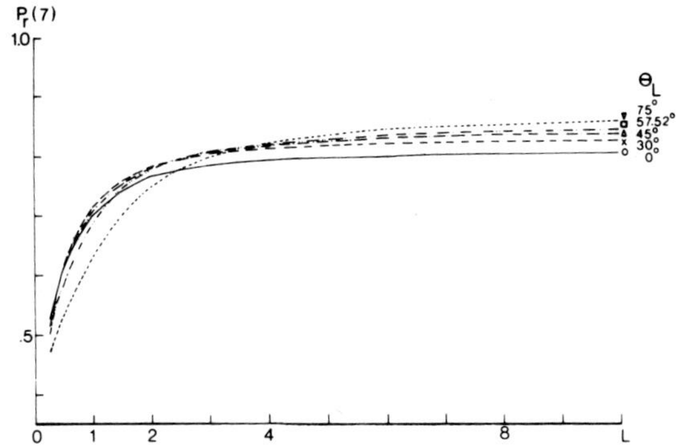


FIG. 4.9 The 'vegetation index' (VI = $P_r(7)$) as a function of LAI (L) for a wheat crop with different leaf angles ($\theta_L$) on a dry sandy loam soil (from: Bunnik, 1978).

In order to find an index independent of soil influence Richardson & Wiegand (1977) introduced the 'perpendicular vegetation index' (PVI):

$$PVI = \sqrt{(MSS5_S - MSS5_V)^2 + (MSS7_S - MSS7_V)^2} \tag{4.8}$$

with s = soil
v = vegetation.

MSS5 and MSS7 are the reflectance values in the red and the infrared Landsat bands, respectively, expressed as digital counts. They found that MSS digital data for bare soil followed a highly predictable linear relation (soil background line) for MSS bands 5 and 7. Increasing vegetation development was associated with displacement of MSS digital counts perpendicularly away from the soil background line (hence 'perpendicular'). Consequently, Richardson & Wiegand applied the perpendicular distance of an MSS measurement from the soil background line as an index of plant development. The PVI has also been used by e.g. Curran (1983), Jackson et al. (1983a).

However, in order to apply the PVI the reflectance of the soil has to be known, and often it is not. Furthermore, often no bare soil is visible within a recording area and the reflectance of the soil cannot be estimated.

A similar approach for suppressing variations in soil background was developed by Kauth & Thomas (1976). They applied a heuristic linear transformation in the four-dimensional data space provided by Landsat MSS measurements for different soils. They extracted a 'soil brightness index' and a 'greenness index' as the main uncorrelated compounds. The transformations for their original set of data from Landsat-1 are:

$$\text{Soil brightness} = 0.43\ MSS4 + 0.63\ MSS5 + 0.59\ MSS6 + 0.26\ MSS7 \tag{4.9}$$

$$\text{Greenness} = -0.29\,\text{MSS4} - 0.56\,\text{MSS5} + 0.60\,\text{MSS6} + 0.49\,\text{MSS7} \tag{4.10}$$

Soil brightness varies strongly with soils; greenness is uncorrelated with soil brightness and varies with different green vegetations. The concept of brightness and greenness has been used by many other authors, e.g. Barnett & Thompson (1983), Bauer et al. (1980, 1982), Hlavka et al. (1979), Jackson et al. (1983b), Thompson & Wehmanen (1980).

The latter transformation cannot be directly applied to the three passbands used in this report. However, since the literature shows that MSS4 and MSS5 (visible passbands) correlate strongly, as do MSS6 and MSS7 (infrared passbands), the greenness index will be very similar to a weighted difference between an infrared and a visible passband.

The concept of a difference between an infrared and a visible passband was applied by Gray & McCrary (1981a, 1981b) using data from NOAA satellites. These satellites only yield recordings in one passband in the visible and in one passband in the near-infrared. Tucker (1979) also suggested using this index to monitor green vegetation and this index in general yielded the best results of all the indices he investigated. However, Tucker excluded the infrared-red difference from further considerations because it will not compensate for differences in irradiation conditions.

If the spectral measurements can be corrected for differences in irradiation in some other way (e.g. by directly measuring or by calculating reflectances), the infrared-red difference will only slightly be influenced by red reflectance after complete soil cover has been attained. So the infrared-red difference may be a feasible estimator for LAI and dry matter weight inasmuch as it compensates to a greater or lesser degree for the influence of soil reflectance (cf. greenness index of Kauth & Thomas, 1976). In the next chapter it will be shown that the infrared-red difference for some soil types can be reasonably satisfactorily used to correct the influence of soil reflectance when used as an estimator of LAI. None of the authors who used this difference deduced this index from any physical reflectance model. Therefore, the mathematical description of the relationship of such an index to crop characteristics such as LAI differs from author to author, being derived in an empirical way. Moreover, researchers have never laid much emphasis on this particular index as being a very powerful index for monitoring vegetation.

Hence, a study of the literature did not reveal an index that is suitable at complete soil cover for discriminating between large LAI values (up to LAI 8). Therefore, in the following section some more complicated models for estimating crop characteristics will be reviewed.

### 4.7.1 Survey of literature

Canopy-modelling studies also enable relationships between reflectance values and crop characteristics to be studied. The main aim of physical reflectance models suitable for agricultural crops is to obtain a better understanding of the complex interaction between solar radiation and plant canopies. Such reflectance models describe the radiation transfer within a canopy or within canopy layers and finally give the total reflectance of a certain vegetation. Essentially, there are two classes of physical reflectance models: numerical models and analytical models. Bunnik (1984) has reviewed several models.

An example of a numerical model has been described by Idso & De Wit (1970). In this model, radiative transfer is determined by scattering and absorption for discrete leaf layers. Scattering involves both the process of reflection and refraction of radiation at the interface of two media. Goudriaan (1977) improved and extended this model by calculating a numerical solution for upward and downward diffuse fluxes within nine sectors of each hemisphere for each discrete layer.

One of the earliest analytical models was described by Allen & Richardson (1968). It was based on a theory of Kubelka & Munk (1931) which describes the transfer of isotropic diffuse flux in perfectly diffusing media. In the analytical model, upward and downward fluxes are expressed by differential equations. Allen et al. (1969) extended this model in order to include scattering of direct solar flux by using the Duntley equations (Duntley, 1942). The first analytical model incorporating both illumination and observation geometry was developed by Suits (1972a) and is an extension of the model developed by Allen and his colleagues. Suits's model also incorporates plant canopy structural (with a drastic simplification) and optical properties. Contrary to the numerical models, the Suits model has been vigorously verified with practical reflectance measurements taken under different circumstances (Chance, 1977; Chance & LeMaster, 1977; Colwell, 1974; LeMaster, 1975; Suits, 1972a,b; Suits & Safir, 1972). Model calculations have agreed satisfactorily with measured reflectances.

Since the physical relations between reflectance and crop characteristics can better be understood with analytical expressions, and a drastic simplification of canopy structure still results in a good correspondence with measured data, an extension of the Suits model (namely the SAIL model, which was available for comparison) will be used for comparing some new model derivations in chapter 5. Therefore the Suits model will be described in more detail below.

### 4.7.2 Suits model

The Suits model is concerned with the interception of solar radiation by a certain canopy and the subsequent scattering of the radiation towards a sensor, with the canopy geometry playing an important role. As is customary with many models it is assumed that the canopy is infinitely extended horizontally and that canopy components have a random uniform distribution within each homogeneous layer. The leaf distribution is described in terms of horizontal (H) and

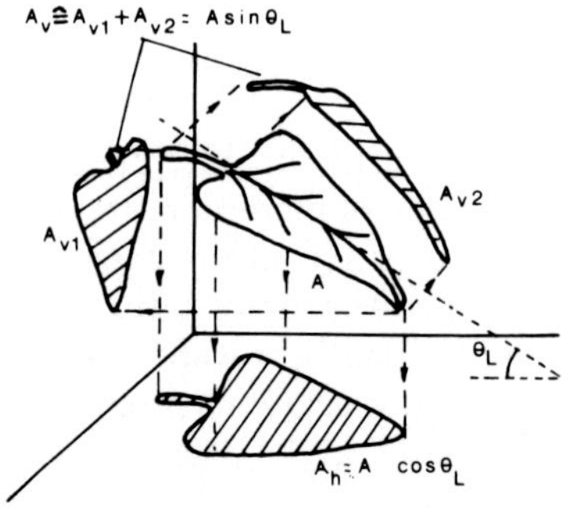


FIG. 4.10 The three orthogonal projections of a canopy component used in the Suits model (after: Bunnik, 1978).

vertical (V) leaf projections, and it is assumed that the only canopy components are small and flat leaves. The projections of a component on horizontal and vertical planes define panel areas that intercept about the same amount of radiant flux as does the original component (as an approximation). These projections are used for calculating the reflectance and extinction of radiant flux for different canopy layers. In this context, extinction is defined as the amount of radiance (or flux) that does not pass through a layer, relative to the amount of radiance (or flux) incident on that layer.

The orthogonal projections of a leaf used by Suits are shown in figure 4.10. For a canopy with a certain LAI the projections will be described as:

$$H = L \cdot \cos\theta_l \tag{4.11}$$

$$V = L \cdot \sin\theta_l \tag{4.12}$$

with

H = total horizontally projected leaf area per unit soil area

V = total vertically projected leaf area per unit soil area, defined as the sum of the projections on the two vertical planes in figure 4.10

$\theta_l$ = effective canopy leaf inclination angle with the horizontal

and

$$L = \sqrt{(H^2 + V^2)} \tag{4.13}$$

L offers a good approximation of the actual LAI and the effective leaf angle is a good approximation of the average leaf angle (Verhoef & Bunnik, 1975).

The horizontally projected area of vertical components in the direction of the sun equals $V.\mathrm{tg}\theta_s$ (figure 4.11), $\theta_s$ being the solar zenith angle.

The model assumes that the projections of leaf normals have a uniform distribution in azimuth. The probability density function equals (with $\phi$ = azimuth angle between leaf normal and sun): $f(\phi) = 1/\pi$. Then the azimuthal average of the fractional area of the vertical components equals:

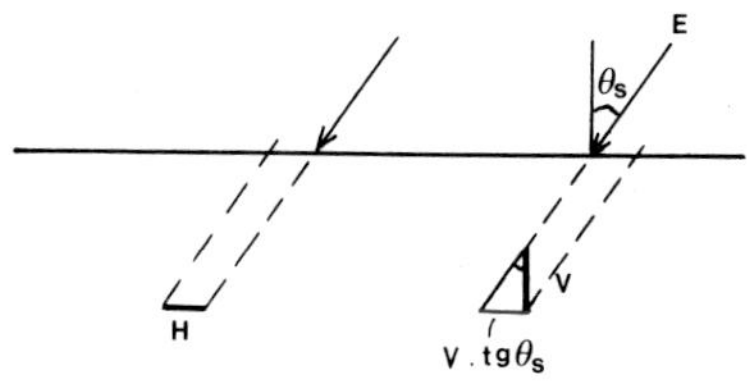


FIG. 4.11 The projection of a vertical component in the direction of the sun.

$$\int_0^{\pi} \sin\phi \cdot f(\phi) \cdot d\phi = \frac{2}{\pi} \qquad (4.14)$$

The extinction coefficient k, referring to extinction of radiation coming from the sun, is now defined as:

$$k = H + \frac{2}{\pi} \cdot V \cdot tg\theta_s \qquad (4.15)$$

This means that only those elements are included that may intercept radiation from the sun. Here one includes all horizontal elements (H) as well as the area of vertical elements projected towards the sun ($V.tg\theta_S$), taking the uniform azimuthal distribution into account (factor $2/\pi$). In the same way the extinction coefficient K in the direction of the sensor is defined as:

$$K = H + \frac{2}{\pi} \cdot V \cdot tg\theta_o \qquad (4.16)$$

with $\theta_0$ = zenith observation angle.

These two coefficients are the key coefficients in the Suits model in that they determine the flux reaching the soil and the flux leaving the canopy in the direction of the sensor. Thus these coefficients are important for estimating soil cover (chapter 5) and subsequently for estimating LAI.

Bunnik (1978) described an approximation of the canopy reflectance in the red for the one-layer case (the diffuse radiant flux is ignored):

$$r \simeq r_\infty + (r_s - r_\infty) \cdot e^{-L \cdot (2\cos\theta_l + \frac{2}{\pi} \sin\theta_l \cdot (tg\theta_o + tg\theta_s))} \qquad (4.17)$$

$r_\infty$ = reflectance of an infinitely thick canopy
$r_s$ = soil reflectance.

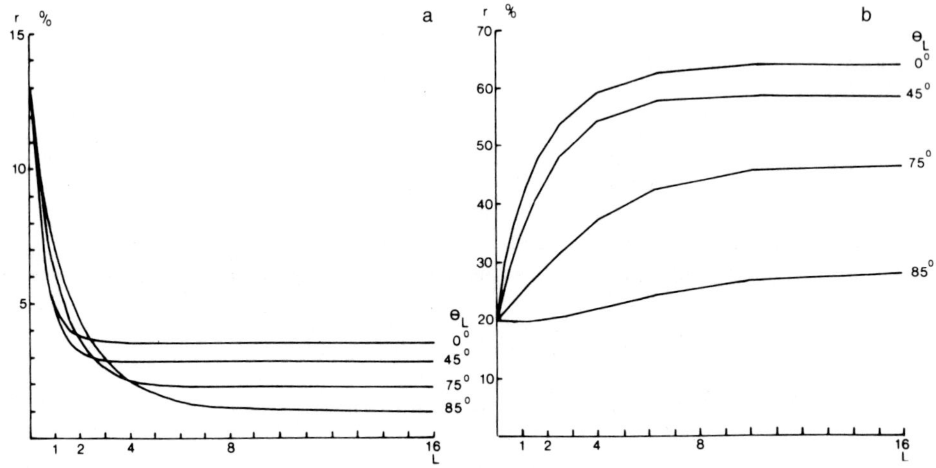


FIG. 4.12 Crop (+ soil) reflectance in (a) the red at 670 nm and (b) the infrared at 870 nm as a function of LAI (L) with different leaf angles ($\theta_L$) (from: Bunnik, 1978).

The exponential function in equation (4.17) describes the relative contribution of the soil. As will be seen in chapter 5, this function coincides with the fraction of the soil visible to the sensor as well as directly illuminated by the sun. This exponential relationship between reflectance in a red passband and LAI is illustrated in figure 4.12. Because of the high transparency of the leaves in the near infrared wavelength region, the relationship between infrared reflectance and LAI is much more complicated (this relationship has also been described by Bunnik (1978)). It is also illustrated in figure 4.12.

In agronomy, soil cover is defined as the relative vertical projection of all plant elements. According to this definition, Bunnik (1978) defined soil cover (B) in the following way:

$$B = 1 - e^{-L \cdot \cos\theta_l} \tag{4.18}$$

Using this definition, Bunnik obtains relationships between reflectance and soil cover, shown in figure 4.13 for a red and an infrared wavelength band.

Suits's model was further extended by Verhoef & Bunnik (1976) to include crop row effects.

Bunnik (1978) compared the Suits model with Goudriaan's model and found similar results.

When model simulations are carried out with varying view angle, Suits's simplifications appear to be too drastic (Verhoef & Bunnik, 1981). Therefore, Verhoef (1984) extended the Suits model further by including scattering and extinction functions for canopy layers containing fractions of oblique leaves (inclined leaves). He did not introduce the drastic simplification of canopy geometry to exclusively horizontal and vertical components as used by Suits, but he used a discretized set of frequencies at distinct leaf angles. The same assumptions

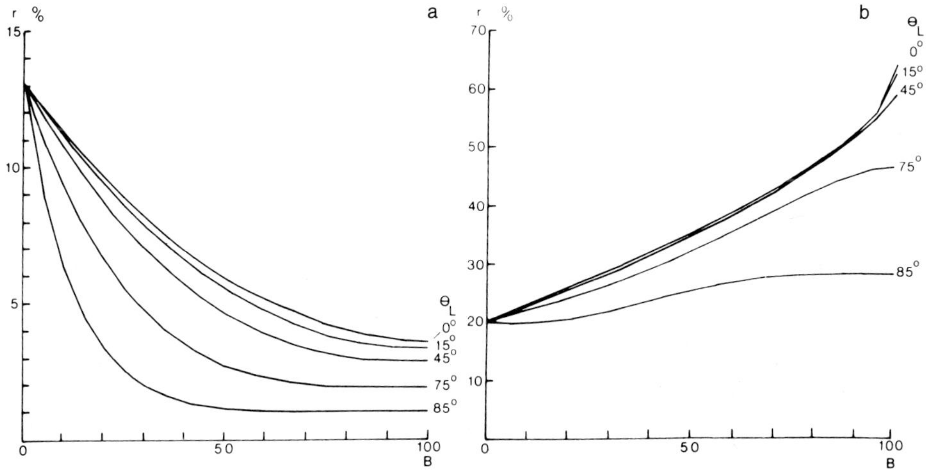


FIG. 4.13 Crop (+ soil) reflectance in (a) the red at 670 nm and (b) the infrared at 870 nm as a function of soil cover (B) with different leaf angles ($\theta_L$) (from: Bunnik, 1978).

are used as with the Suits model, and it contains the Suits model as a special case. This model is called the SAIL model (Scattering by Arbitrarily Inclined Leaves).

All models discussed so far are based on the assumption that the canopy is infinitely extended in the horizontal plane. Kimes & Kirchner (1982) have developed a three-dimensional model, which describes the radiative transfer for heterogeneous scenes by subdividing the scene into modules. Long computing times are needed for calculations with this model.

## 4.8 Summary

This chapter begins by giving definitions of reflection and spectral measurements in general, the main spectral characteristic in the present study being the reflectance (or reflectance factor).

Reflectance values in the visible region of the electromagnetic spectrum may be most appropriate for estimating soil cover by vegetation. At complete cover the visible reflectance may provide information about leaf colour. Reflectance values in the infrared may be the most appropriate for estimating LAI. Other plant characteristics, such as yield, diseases and lodging may be estimated indirectly from reflectance values. Yield, for instance, often correlates strongly with LAI. The possibilities of making such estimations will be analysed in chapter 7, using measurements acquired by the remote sensing system described in chapter 3.

At low soil cover, soil reflectance will contribute largely to the measured signal. Differences in soil moisture content will greatly influence soil reflectance.

Thus, in ascertaining the relationship between reflectance and crop characteristics, a correction for soil reflectance should be made, especially in a multitemporal analysis (cf. chapter 5).

In published work, indices are often used for estimating crop characteristics, in this case LAI. In this report, reflectances are always calculated and therefore the infrared-red difference as an index may be a feasible estimator of LAI; this will be shown in chapter 5. In other work often some ratio of reflectances is preferred, because one is unable to calculate reflectances accurately and a ratio will correct for differences in illumination to some extent. Also, more complicated reflectance models may be used for predicting LAI. Such models usually simulate reflectances for varying crop characteristics and incorporate simplified structural and optical properties of the canopy. Although for practical applications a simple function of reflectances for estimating e.g. soil cover or LAI is preferred, a physical basis is unavoidable in order to deduce the kind of relationship between such a function and soil cover or LAI. Therefore, in the next chapter a new model will be introduced that results in simple functions for estimating soil cover and LAI and incorporates a correction for soil background. Mathematical relationships between these functions and crop characteristic are described. These will be verified by means of calculations with the SAIL model.

# 5 SIMPLIFIED REFLECTANCE MODEL FOR VEGETATION

## 5.1 INTRODUCTION

As discussed in the previous chapter, in the literature no index (which is a simple function of reflectances) stands out as being particularly suitable for estimating crop characteristics (such as LAI) in agricultural field trials. Many indices lose sensitivity, especially at large LAI values (above 3–4). Moreover, no mathematical description of a physical relationship between indices and crop characteristics has been found. On the other hand, published reflectance models require many input parameters, some of which are difficult to ascertain (e.g. leaf angle distribution). This reduces the usefulness of these models for the agronomist. Since such models also calculate reflectances from crop characteristics, they would have to be inverted so that they could be used to calculate crop characteristics from reflectances. This is very difficult, if not impossible, for most models.

Therefore, for the present study a simplified reflectance model for vegetation had to be created. In this chapter such a model will be introduced. The main requirements for such a model are:

1. it should be possible to estimate LAI (which is the main crop characteristic currently used in agronomy);
2. it should describe the relationship between reflectance and LAI by more or less physically defined parameters;
3. it should correct for soil background in order to enable a multitemporal analysis to be done;
4. it should be as simple as possible (preferably resulting in some sort of index).

## 5.2 MODEL PRESENTATION

As described in chapter 4, soil cover is estimated best from reflectance values in visible passbands, because the transmission of radiance through leaves is almost negligible in these passbands. LAI, however, is estimated best from reflectance values in an infrared passband.

Conventionally, for a green canopy, soil cover is defined as the vertical projection of the canopy on the soil surface (figure 5.1). In practice, however, the researcher will estimate the relative amount of soil visible to him. When one looks vertically downwards at a well-developed wheat crop, for instance, some shadows may be visible between the stems of the plants, and when looking from a height of a few metres or more above the top of the canopy (a realistic situation for remote sensing) one is unable to decide whether there are leaves present in the shadows just above the soil. In a situation where no soil is actually visible

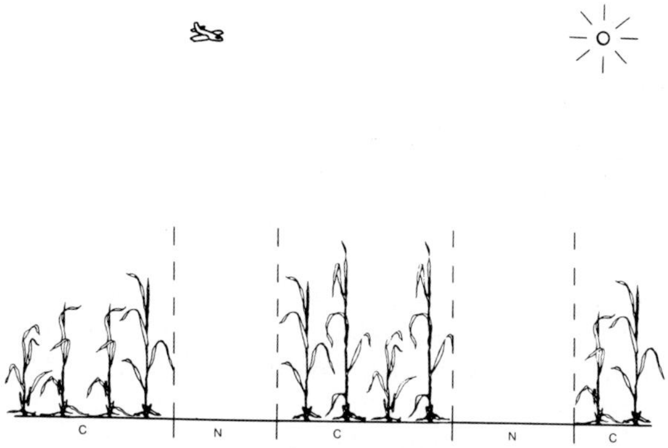


FIG. 5.1 Schematic presentation to illustrate the conventional definition of soil cover.
C = covered soil
N = soil not covered

to the researcher, he will judge that soil cover is complete. However, this same situation may be judged to be an incompletely covered soil if the original definition referring to the vertical projection of the canopy is strictly applied.

If a remote sensing technique is used in a visible passband while looking downwards from some distance, the sensor will also be unable to detect whether soil in the shadows is obscured by leaves. For this reason the conventional definition of soil cover is not satisfactory and therefore I propose to redefine soil cover, giving a new definition more appropriate for remote sensing.

A prerequisite for ascertaining bare soil according to this new definition is that the soil must clearly contrast with the vegetation. When the sun is shining the soil should be directly illuminated by the sun (figure 5.2a). If the soil is not directly illuminated, it will be impossible to say whether it is obscured by green vegetation or by shadows. Both have a low reflectance value in the visible region of the electromagnetic spectrum, with little contrast. Further, the soil must be visible for the detector to be able to classify it as soil (figure 5.2b). The fraction of soil that satisfies both conditions (i.e. soil that is illuminated by the sun as well as directly detectable by the sensor) will be classified as the fraction of soil that is not covered (figure 5.2c). The complementary fraction will now be defined as soil cover. In the special situation of the sensor looking vertically downwards (as will be predominantly the case in the investigations in this monograph), this definition of soil cover is equivalent to the relative vertical projection of green vegetation, the relative area of the shadows included.

If there is only indirect sunlight (overcast sky), which is called skylight, there will be a gradual transition in detected signal from brighter soil at some distance from the plants, to darker soil near the plants. Under such circumstances, soil that is not covered is not as clearly distinguishable as with direct sunlight. Nevertheless, the same approach may be applied for estimating an 'apparent' soil cover. Such an estimated soil cover may also be used for calculations in an infra-

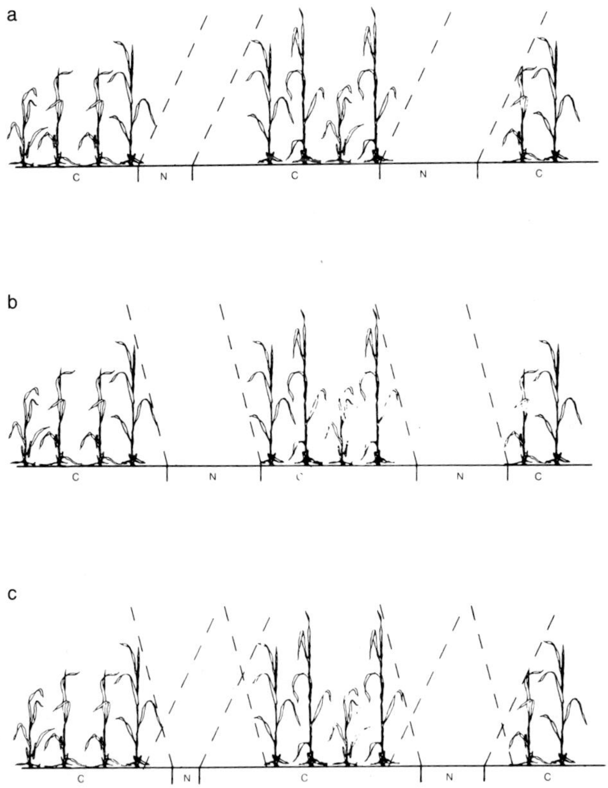


FIG. 5.2 Schematic presentation to illustrate aspects of the new definition of soil cover.
a: Soil illuminated by the sun.
b: Soil visible to a sensor.
c: Illuminated soil visible to this sensor.

red passband (cf. section 5.6), because shadow in the visible region will also be detected as shadow in the infrared region (cf. BUNNIK & VERHOEF, 1974). Since most remote sensing recordings in the present study were obtained with direct sunlight (clear sky), only this situation will be elaborated further.

A major problem in estimating soil cover from reflectances is soil moisture content. In section 5.5 the ways that differences in soil moisture content can be corrected for will be discussed.

We decided to estimate LAI in the agricultural field trials from reflectance values in an infrared passband, because transmission here is not negligible and the underlying leaves contribute significantly to total reflectance. At the beginning of the growing season, total reflectance will largely be determined by soil reflectance. Soil reflectance, however, is not constant during the season, because of changes in soil moisture content. Consequently, comparison of the infrared reflectances on different dates cannot be justified. Yellow and dead leaves become more common towards the end of the growing season, and since we are primarily interested in the reflectance of green leaves (cf. definition LAI), this is another reason why comparison of infrared reflectances from different dates is inadvisable. In order to ascertain whether there is a useful relationship between infrared reflectance and LAI, the former should be corrected for background (soil, or yellow and dead leaves insofar as they are visible), because it may influence infrared reflectance independently of the LAI. The infrared reflectance is then calculated for the situation of the visible background being completely black and not reflecting any radiation. This corrected infrared reflectance value is then used to estimate LAI.

Let us consider the simple situation of a surface, partly covered with vegetation and partly bare (figure 5.3). The fraction of the surface covered with vegetation is called soil cover, B. If the reflectance of the soil is called $r_s$ and the reflectance of the vegetation $r_v$, then the total measured reflectance, r, will equal:

$$r = r_v \cdot B + r_s \cdot (1-B) \qquad (5.1)$$

An analogous equation has been reported by other researchers, e.g. Badhwar (1980), Goillot (1980), Jackson et al. (1979) and Rao et al. (1979).

A green passband will be denoted by the subscript g and equation (5.1.) is then written as:

$$r_g = r_{v,g} \cdot B + r_{s,g} \cdot (1-B) \qquad (5.2)$$

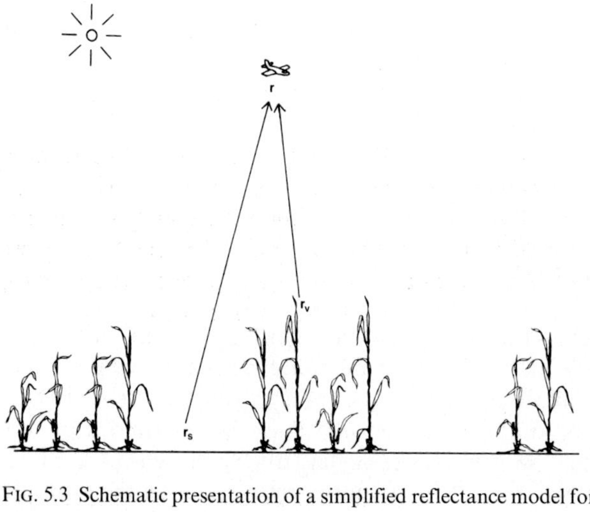


FIG. 5.3 Schematic presentation of a simplified reflectance model for vegetation and soil combined.

The reflectance of vegetation in a green passband ($r_{v,g}$) may be regarded as being independent of the number of leaf layers, because leaf transmittance in the green is assumed to be negligible. Hence equation (5.2) describes the linear relationship between the reflectance in a green passband and soil cover, if the soil reflectance can be considered to be constant (constant soil moisture content).

Analogously, by attaching the subscript r to a red passband reflectance, we have:

$$r_r = r_{v,r} \cdot B + r_{s,r} \cdot (1 - B) \qquad (5.3)$$

Because $r_{v,r}$ can also be regarded as being independent of the number of leaf layers, this equation describes the linear relationship between red reflectance and soil cover.

For an infrared passband the subscript ir will be used and equation (5.1) is then written as:

$$r_{ir} = r_{v,ir} \cdot B + r_{s,ir} \cdot (1 - B) \qquad (5.4)$$

In this equation $r_{v,ir}$ is not independent of the number of leaf layers, so it may not be regarded as a constant. It is composed of a single scattering component and a multiple scattering component. The latter component is not a constant, but depends (among other things) on the number of leaf layers. This means that the reflectance measured in an infrared passband ($r_{ir}$) is not a linear function of soil cover.

In order to deduce the relationship between soil cover and LAI, the process of extinction of radiation in a canopy should be considered. If a canopy has a certain extinction coefficient (cf. section 4.7) per leaf layer as well as a certain LAI (abbreviated as L in the formulae), the product of both factors equals the mean extinction of that canopy. The mean extinction consists of two components:

1. extinction in the direction of the sensor, indicated by K.L, where K is the extinction coefficient per leaf layer in the sensor direction;
2. extinction in the sun's direction, indicated by k.L, where k is the extinction coefficient per leaf layer in the sun's direction.

Consider the process of extinction in a very small part (or element) of the canopy. In the visible passbands, extinction in an element occurs when a leaf is hit by radiation. The probability of hitting i elements among n independent elements has a binomial distribution. If the number n of independent elements increases to infinity while the probability of hitting a specific element decreases to zero, the binomial distribution can be approximated by a Poisson distribution. The Poisson distribution states:

$$P(\underline{x} = i) = \frac{e^{-\lambda} \cdot \lambda^i}{i!} \qquad i = 0,1,2,3,.... \qquad (5.5)$$

The random variable x is the number of independent elements of the canopy in which extinction occurs; $\lambda$ is the mean or expected number of elements in which extinction occurs. The probability that no element is hit $(i=0)$ equals:

$$P(\underline{x}=0) = e^{-\lambda} \quad (5.6)$$

The probability of soil being visible in the direction of the sensor equals:

$$e^{-K.L}.$$

The probability of soil being illuminated by the sun equals:

$$e^{-k.L}$$

If one assumes both events to be independent, the probability of sensing illuminated soil equals:

$$e^{-K.L} * e^{-k.L} = e^{-(K+k).L}.$$

The complementary probability is equal to soil cover (new definition). This means that soil cover may be described as:

$$B = 1 - e^{-(K+k).L} \quad (5.7)$$

For the special situation where the zenith and azimuth angles of sensor and sun are equal (hot-spot situation) the probability of sensing illuminated soil equals one. With nadir (vertical) observation the hot-spot situation is comparable with the situation of the conventional definition of soil cover, in which the position of the sun is irrelevant. Then soil cover equals:

$$B = 1 - e^{-K.L} \quad (5.8)$$

For estimation of LAI the corrected reflectance r′, defined as $r' = r - r_s \cdot (1-B)$, could be used. The corrected reflectance is the reflectance one would have obtained with a black background. Because of equation (5.1) we have:

$$r' = r_v \cdot B \quad (5.9)$$

Inserting equation (5.7) into (5.9) gives:

$$r' = r_v \cdot (1 - e^{-(K+k) \cdot L}) \quad (5.10)$$

So, in principle, it is possible to estimate the LAI by means of reflectance measurements in a visible passband. However, the corrected reflectance (r′) already tends to the asymptotic value ($r_v$) when the LAI value is low, because $K+k$ are relatively large (Goudriaan, 1977).

In an infrared passband the corrected reflectance as a function of LAI approaches an asymptotic value much more slowly (section 4.7), because $r_v$ in equation (5.10) is not a constant, but is a complicated function of, among other things, the scattering properties of the canopy. Thus, the infrared reflectance can be more suitable than the visible reflectance for estimating LAI.

Bunnik (1978) has indicated that the relationship between infrared reflectance and LAI can be described by means of a linear combination of several exponential functions in which canopy structure as well as the scattering properties of the canopy are incorporated. Let us investigate whether this relationship can be described with only one exponential function. The relationship between LAI and infrared reflectance (cf. figure 4.12) greatly resembles a 'Mitscherlich curve' (y = A-b.exp(-k.t)). Such a curve has 3 parameters. In the special situation of such a curve running through the origin (A = b), the curve defined by y = A.(1-exp(-k.t)) has only 2 parameters. For describing the relationship between the corrected infrared reflectance and LAI (which runs through the origin) an empirical equation similar to equation (5.10) could be used:

$$r'_{ir} = r_{\infty,ir} \cdot (1 - e^{-\alpha . L}) \qquad (5.11)$$

parameter $r_{\infty,ir}$ being the asymptotic value for the infrared reflectance and parameter $\alpha$ a combination of extinction and scattering coefficients. This is the special case of a Mitscherlich curve.

From equation (5.11) it follows that if LAI = 0 then $r'_{ir} = 0$
and if LAI $\rightarrow \infty$ then $r'_{ir} = r_{\infty,ir}$.

Finally the LAI is solved from equation (5.11):

$$L = -1/\alpha \cdot \ln(1 - r'_{ir}/r_{\infty,ir}) \qquad (5.12)$$

## 5.3 Relationship between the model developed above and the Suits model

Combining equations (4.11), (4.12), (4.15) and (4.16) from the Suits model with equation (4.17), which Bunnik (1978) deduced from the Suits model for directional reflectance with single scattering, results in the equation:

$$r = r_{\infty} + (r_S - r_{\infty}) \cdot e^{-K-k} = r_{\infty} \cdot (1 - e^{-K-k}) + r_S \cdot e^{-K-k} \qquad (5.13)$$

where $r_{\infty}$ is the asymptotic value of the reflectance factor of vegetation in a visible passband. In Bunnik's notation K (or k) is the product of the extinction coefficient and the LAI, and therefore in the context of the present report K (or k) should be replaced by K.L (or k.L).

$$r = r_{\infty} \cdot (1 - e^{(-K-k) \cdot L}) + r_S \cdot e^{(-K-k) \cdot L} \qquad (5.14)$$

In combination with equation (5.7) this gives:

$$r = r_{\infty} \cdot B + r_S \cdot (1 - B) = r_S + (r_{\infty} - r_S) \cdot B \qquad (5.15)$$

For a visible passband this equation equals equation (5.1), because $r_{\infty}$ equals $r_V$ and is constant. For an infrared passband, equation (5.15) is only valid if the contribution of the multiple scattering component is negligible. If this latter term is of importance equation (5.4) should be used ($r_{v,ir} \neq r_{\infty,ir}$).

## 5.4 A DIGRESSION ABOUT INACCURACY

In much remote sensing research work in which reflectance values are used for estimating crop characteristics, a function of two or more reflectance values is used in order to correct for some kind of disturbance (section 4.6). This function is often simply the difference between, or ratio of, two passbands. The inaccuracies involved in such a function are often ignored.

The inaccuracy can be indicated by the standard deviation ($\sigma$) or by the coefficient of variation ($CV = \sigma/\mu$ with $\mu$ = mean). Assume that reflectances in two passbands are represented by $\underline{x}$ and $\underline{y}$ and that $\underline{x}$ and $\underline{y}$ are stochastically independent and have the same CV.

If $\underline{x}$ and $\underline{y}$ are independent, the variance of the difference between $\underline{x}$ and $\underline{y}$ equals the sum of the variances of $\underline{x}$ and $\underline{y}$:

$$\text{var}(\underline{x}-\underline{y}) = \text{var}(\underline{x}) + \text{var}(\underline{y}).$$

For the difference between $\underline{x}$ and $\underline{y}$, two situations may be distinguished:

1. $\underline{x}$ is much larger than $\underline{y}$:
   if $\varepsilon\underline{x} \gg \varepsilon\underline{y}$ then $\varepsilon(\underline{x}\text{-}\underline{y}) \approx \varepsilon\underline{x}$
   and $\text{var}(\underline{x}\text{-}\underline{y}) \approx \text{var}(\underline{x})$
   $\rightarrow CV(\underline{x}\text{-}\underline{y}) \approx CV(\underline{x})$, which is not extremely inaccurate compared with the inaccuracy of $\underline{x}$ and $\underline{y}$.

   (E.g.: x = infrared reflectance of 50%; y = red reflectance of 2%. Assume that both are ascertained independently, each with CV = 0.10. Then the difference is 48%, and the CV = 0.10).

2. $\underline{x}$ and $\underline{y}$ are nearly equal:
   if $\varepsilon\underline{x} \approx \varepsilon\underline{y}$ then $\varepsilon(\underline{x}\text{-}\underline{y}) \approx 0$
   and $\text{var}(\underline{x}\text{-}\underline{y}) \approx 2.\text{var}(\underline{x})$
   $\rightarrow CV(\underline{x}\text{-}\underline{y})$ will be very large, and the inaccuracy of the difference can be extremely large.

   (E.g.: x = red reflectance of 12%; y = green reflectance of 11%. Assume that both are ascertained independently, each with CV = 0.10. Then the difference is 1%, and the CV = 1.63).

If $\underline{x}$ is independent of $\underline{y}$, the square of the CV value of the ratio of $\underline{x}$ to $\underline{y}$ is approximately the sum of the squares of the CV values of $\underline{x}$ and $\underline{y}$:

$$CV^2\left(\frac{\underline{x}}{\underline{y}}\right) \approx CV^2(\underline{x}) + CV^2(\underline{y})$$

If the CV of $\underline{x}$ equals the CV of $\underline{y}$ then the CV value of the ratio is equal to the square root of 2, times the CV value of $\underline{x}$ or $\underline{y}$.

(E.g.: x = red reflectance of 12%; y = green reflectance of 11%. Let us assume that both x and y are ascertained independently, each with CV = 0.10. Then the ratio is 1.09, and the CV = 0.14).

When reflectances are measured from an airborne or spaceborne platform, a relatively large CV value may occur in one passband if the expected value is very small (e.g. red reflectance). This may be caused by a small signal-to-noise ratio. If the reflectance of an object is very small, the influence of the atmosphere on the detected signal will be large. As a result, inaccuracies in correcting for these atmospheric effects will have a large effect on the resulting inaccuracy of the derived reflectance of the object. (E.g.: the sensor detects a reflectance of 7%: 2% is the reflectance from the object and 5% is the contribution of the atmosphere. Assume that the values 7% and 5% are ascertained independently, each with CV = 0.10. Then the difference, 2%, will be the reflectance of the object and the CV will be 0.43). Then the ratio of two passbands may also result in a very inaccurate value.

Thus, it can be concluded that one has to be careful in using a difference or ratio of two passband measurements, as in section 4.6. A single passband is often preferable, if its inaccuracy is not too large. If the values in two passbands differ considerably, the difference between two passbands will be preferable. If the values in two passbands are nearly the same, the ratio between the measurements in two passbands will be preferable.

## 5.5 Estimation of soil cover

As stated before (section 5.2) reflectance measurements in a visible passband are very suitable for estimating soil cover. Equation (5.1) may be rewritten as:

$$B = \frac{r_s - r}{r_s - r_v} \tag{5.16}$$

For a visible passband this equation describes the linear relationship between soil cover and the measured reflectance. This equation may be applied to a green passband or to a red passband. However, the reflectance of bare soil ($r_S$) and of vegetation ($r_V$) should be known. Although it is quite often possible to ascertain a good estimate for the reflectance of vegetation (complete cover), estimating the reflectance of bare soil poses greater difficulties. The reflectance of a soil may change very rapidly, according to soil moisture content. Also, very large local differences in soil moisture content may occur. Soil moisture content is an important factor in determining the relationship between reflectance and soil cover, as is illustrated in figure 5.4. From figure 5.4 it is evident that if soil moisture content is not known, huge errors may be made in estimating soil cover. Three methods that may enable this problem to be overcome are described below.

For many soil types, reflectance in the different passbands does not differ very much (e.g. Condit, 1970); often there is a slight increase in reflectance with increasing wavelength.

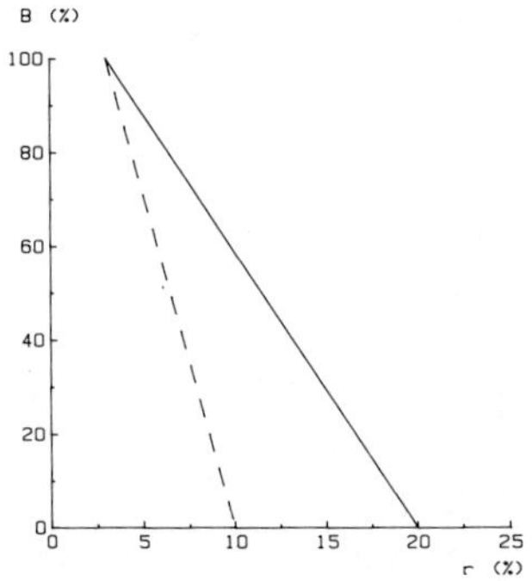


FIG. 5.4 Model calculation of soil cover (B) as a function of reflectance (r) in a visible passband for a dry (—; $r_s = 20.0$) and a wet (---; $r_s = 10.0$) soil, while $r_v = 3.0$.

Let us assume that there is a known linear relationship between the reflectance factors of soil in two passbands (e.g. green and red):

$$r_{s,g} = a + b \cdot r_{s,r} \qquad (5.17)$$

where a and b are two parameters.

Combining equation (5.17) with equations (5.2) and (5.3) gives:

$$r_g - b \cdot r_r = (r_{v,g} - b \cdot r_{v,r}) \cdot B + a \cdot (1 - B) \qquad (5.18)$$

METHOD 1:

As stated before, many soil types have a constant ratio of green and red reflectance of bare soil, which is independent of soil moisture content.

$$\frac{r_{s,g}}{r_{s,r}} = C_1 \qquad (5.19)$$

With regard to equation (5.18) this means a = 0 and b = $C_1$. If we assume that we are able to determine this constant $C_1$, merely by measuring both reflectance values at the same soil moisture content, then equations (5.2), (5.3) and (5.19) offer us 3 equations with 3 unknown variables: the reflectance of bare soil in a green ($r_{s,g}$) and in a red ($r_{s,r}$) passband and soil cover (B). Combining these three equations should give us a solution for estimating soil cover independently of soil moisture content.

$$B = \frac{r_g - C_1 \cdot r_r}{r_{v,g} - C_1 \cdot r_{v,r}} \qquad (5.20)$$

The result of this equation is illustrated in figure 5.5. So, if there is a constant known ratio between the green and red reflectance of bare soil it is possible to estimate soil cover without knowing anything about soil moisture content or the actual soil reflectances.

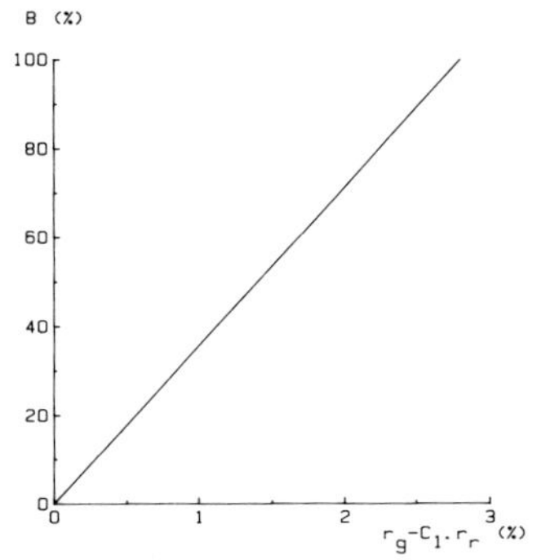


FIG. 5.5 Model calculation of soil cover (B) as a function of $r_g$ - $C_l \cdot r_r$ (method 1) with $r_{v,g}$= 5.0, $r_{v,r}$ = 2.0 and $C_l$ = 1/1.1.

METHOD 2:

If the reflectance of bare soil in the green and red passbands does not differ very much, as is the case for many soil types, the difference between green and red reflectance will be nearly independent of soil moisture content. Combining equations (5.2) and (5.3) gives:

$$B = \frac{(r_{s,g}-r_{s,r}) - (r_g-r_r)}{(r_{s,g}-r_{s,r}) - (r_{v,g}-r_{v,r})} \qquad (5.21)$$

With regard to equation (5.18) this means a = 0 and b≈1. This means that there is a linear relationship between soil cover and the difference between green and red reflectance, and that this relationship is nearly independent of soil moisture content (there is a distinct difference in reflectance between both passbands at complete cover). This is illustrated in figure 5.6. So the difference between green and red reflectance offers another opportunity for estimating soil cover

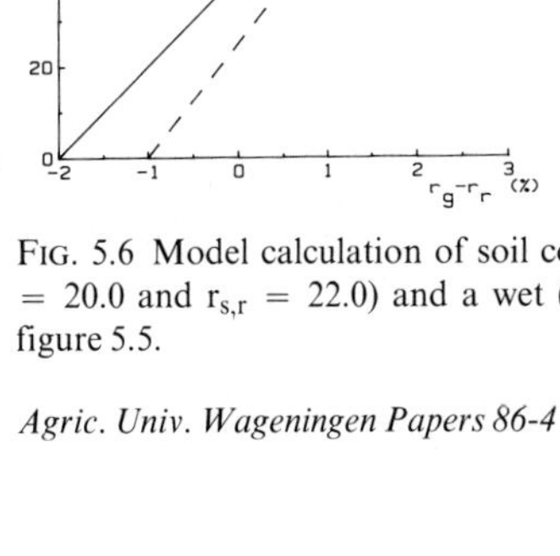


FIG. 5.6 Model calculation of soil cover (B) as a function of $r_g$ - $r_r$ (method 2) for a dry (—; $r_{s,g}$ = 20.0 and $r_{s,r}$ = 22.0) and a wet (---; $r_{s,g}$ = 10.0 and $r_{s,r}$ = 11.0) soil, with $r_{v,g}$ and $r_{v,r}$ as in figure 5.5.

for certain soil types, without any knowledge about soil moisture content. However, as figure 5.6 clearly shows, inaccuracy may be considerable, especially at low soil cover values, if the reflectances of bare soil are not identical.

In all cases the linear relationship between the reflectances in the two passbands has to be tested and ascertained once for the soil type concerned. Afterwards this linear relationship can be used to correct for an unknown change in soil moisture content, in order to estimate soil cover.

When taking the difference between green and red reflectance, e.g. equations (5.20) or (5.21), one has to remember the dangers of inaccuracy (see section 5.4). Generally, the values of green and red reflectance are similar and the difference will be small in absolute value. As seen in section 5.4 this may lead to great inaccuracy.

METHOD 3:

If the ratio of the green and red reflectance of bare soil is independent of soil moisture content, taking the ratio of green and red reflectance values may improve the estimation of soil cover. Combining equations (5.2) and (5.3) gives:

$$B = \frac{r_{s,g} - r_{s,r} \cdot r_g/r_r}{r_g/r_r \cdot (r_{v,r} - r_{s,r}) - (r_{v,g} - r_{s,g})} \tag{5.22}$$

This ratio is no longer a linear function of soil cover. More important, however, is the conclusion that the influence of soil moisture content has not been completely eliminated (cf. Bunnik, 1978). This is illustrated in figure 5.7. However, if one does not know the soil moisture content and the corresponding reflectance of bare soil reasonably well, this ratio enables soil cover to be estimated more accurately than when the information acquired in a single passband is used. Unless the inaccuracy of the absolute reflectance values in the visible passbands is small, a ratio function may also be more preferable for estimating soil cover than a difference function.

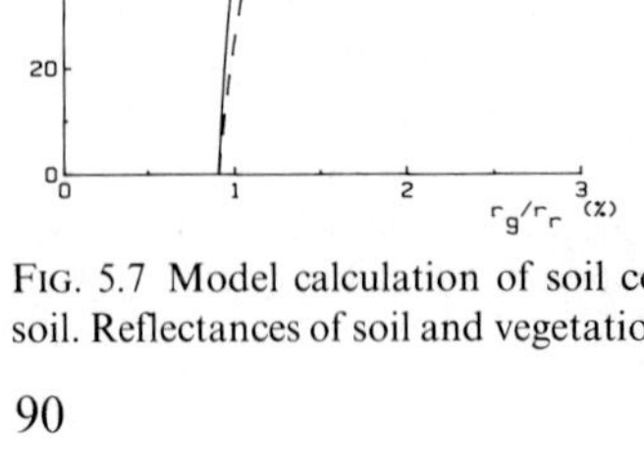


FIG. 5.7 Model calculation of soil cover (B) as a function of $r_g/r_r$ for a dry (—) and a wet (---) soil. Reflectances of soil and vegetation as in figure 5.6.

Reflectance measurements in an infrared passband are most suitable for estimating LAI. LAI is estimated from equation (5.12), which is independent of soil moisture content. However, equation (5.9) first has to be applied to the infrared passband in order to obtain the corrected infrared reflectance:

$$r'_{ir} = r_{ir} - r_{s,ir} \cdot (1 - B) \tag{5.23}$$

Differences in soil reflectance that result from differences in soil moisture content modify this relation in a manner similar to that encountered in estimating soil cover (see section 5.5). This is illustrated in figure 5.8. At low soil cover values this may cause large inaccuracy if neither the soil moisture content nor the actual reflectance of the soil are known. To obtain an accurate estimate of LAI one either has to know or to measure the reflectance of the bare soil, or one has to derive a relation that is less dependent on differences in soil moisture content. Two methods that may enable the problem of not knowing soil moisture content (unknown soil reflectances) to be overcome are described below.

METHOD 1:

Theoretically, it is possible to calculate the corrected infrared reflectance ($r'_{ir}$) accurately, without knowing the soil moisture content or soil reflectance for each individual situation. The only assumption about soil characteristics one has to make is that the ratio between the reflectances in two passbands is constant. As stated in section 5.5, this assumption is valid for many different soil types. This implies that not only equation (5.19) is valid but also:

$$\frac{r_{s,ir}}{r_{s,r}} = C_2 \tag{5.24}$$

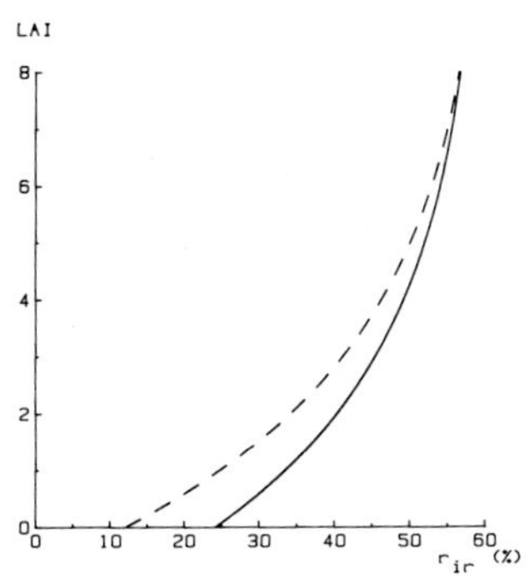


FIG. 5.8 Model calculation of LAI as a function of reflectance in an infrared passband ($r_{ir}$) for a dry (—; $r_{s,ir}$ = 24.2) and a wet (---; $r_{s,ir}$ = 12.1) soil, with $\alpha$ = 0.3 and $r_{\infty,ir}$ = 60. Reflectances of soil and vegetation in visible passbands as in figure 5.6.

Analogously to the situation for estimating soil cover, where we had 3 equations with 3 unknown variables, we are now able to solve the corrected infrared reflectance from 5 equations with 5 unknown variables. These 5 equations are (5.2), (5.3), (5.19), (5.23) and (5.24) and the 5 unknown variables are the corrected infrared reflectance ($r'_{ir}$), soil cover (B) and the soil reflectance in the three passbands ($r_{s,g}$, $r_{s,r}$ and $r_{s,ir}$). After solving these equations for the corrected infrared reflectance we obtain:

$$r'_{ir} = r_{ir} - \frac{C_2 \cdot (r_g \cdot r_{v,r} - r_r \cdot r_{v,g})}{C_1 \cdot r_{v,r} - r_{v,g}} \tag{5.25}$$

If the constant ratios $C_1$ and $C_2$ are known, this equation indicates how the corrected infrared reflectance can be calculated independently of soil moisture content. In the situation of bare soil only, $r_g$, $r_r$ and $r_{ir}$ equal $r_{s,g}$, $r_{s,r}$ and $r_{s,ir}$, respectively, and equation (5.25) results in: $r'_{ir} = 0$. In the situation of complete soil cover, $r_g$ and $r_r$ equal $r_{v,g}$ and $r_{v,r}$, respectively, and equation (5.25) results in: $r'_{ir} = r_{ir}$; in other words no correction for soil background is applied if the soil is not visible.

After correcting for soil background by means of equation (5.25), the LAI is estimated by means of equation (5.12) (figure 5.9).

METHOD 2:

Combination of equations (5.16), applied e.g. to a red passband, and (5.23) gives:

$$r'_{ir} = r_{ir} - r_{s,ir} \cdot \frac{r_r - r_{v,r}}{r_{s,r} - r_{v,r}} \tag{5.26}$$

If the reflectance of bare soil in the red passband ($r_{s,r}$) is large compared with the reflectance of the green vegetation ($r_{v,r}$), this latter reflectance, which is very

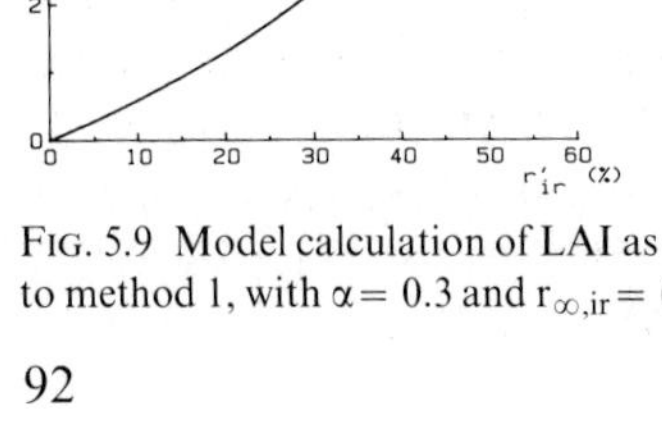


FIG. 5.9 Model calculation of LAI as a function of the corrected infrared reflectance ($r'_{ir}$) according to method 1, with $\alpha = 0.3$ and $r_{\infty,ir} = 62$. Reflectances of soil and vegetation as in figure 5.8.

small, could be omitted from the denominator. If the soil type under consideration has a similar reflectance in the red and infrared passbands, equation (5.26) may be approximated by equation (5.27):

$$r'_{ir} = r_{ir} - r_r + r_{v,r} \tag{5.27}$$

In the situation of bare soil the term $r_{v,r}$ should be omitted in order to get the same result as in equation (5.26) (under the assumption $r_{s,r} = r_{s,ir}$). In the situation of high soil cover the term $r_{v,r}$ is very small compared with ($r_{ir}$–$r_r$), so it may be omitted. A crude approximation for estimating the corrected infrared reflectance will result in the equation:

$$r'_{ir} = r_{ir} - r_r \tag{5.28}$$

To apply this equation for estimating LAI, the difference between the infrared and red reflectance must be ascertained and then equation (5.12) must be used. In this regard $r_{\infty,ir}$ in equation (5.12) will be the asymptotic value of the difference between infrared and red reflectance at very high LAI. If in equation (5.25) the measured reflectances in the green and red passband are assumed to be equal ($r_g = r_r$), then this equation is equal to equation (5.28) under the assumption $C_1 = C_2 = 1$. This drastic approach will be tested in section 5.8 with a data set calculated by means of Verhoef's SAIL model (section 4.7), and provided by him. Furthermore it will be verified with practical data (section 7.6).

When estimating LAI, the considerations raised in section 5.4 should also be taken into account. In contrast to the situation of estimating soil cover, a difference between infrared and visible passbands will give fewer problems. Any such difference could only introduce large inaccuracy if the values in the two passbands (e.g. infrared and red: equation 5.28) are nearly the same: this will occur when soil cover values are low. However, from figure 5.9 it becomes evident that with low soil cover the slope of this curve is not steep; consequently, large inaccuracy in reflectance measurements will introduce only a minor inaccuracy in LAI estimation.

## 5.7 Leaf senescence

At the end of the growing season, annual agricultural plants will show signs of senescence. Leaves turn from green to yellow. This phenomenon starts when the LAI is at its maximum value. In cereals all the leaves have appeared by that moment and the ears are about to appear. Subsequently, both LAI and photosynthetic activity decrease, because only the green parts will be photosynthetically active. During this stage it is important to gain an impression of the speed of senescence and to estimate LAI.

Ascertaining LAI by harvesting plants is a very tedious procedure during senescence. One has to classify leaves as green or yellow: there is no scope for an intermediate category for leaves that are in a transitional stage. However, the photosynthetic activity of a discolouring leaf will be somewhere between the high activity of a green leaf and the zero activity of a dead, yellow leaf. Thus, measuring the LAI during senescence is a rather subjective, inaccurate procedure.

The literature reveals little about the changes in reflectance that occur during senescence. Because leaves change colour, the reflectance in the visible passbands will increase. This means a return to a situation comparable with an increasing contribution from bare soil. However, the ratio between green and red reflectances from a senescing crop may be quite distinct from that from bare soil. Ahlrichs & Bauer (1983) found that the spectral reflectances for a wheat canopy at the seedling and mature stages were similar. After estimating the reflectance of a yellow vegetation, it may be possible to estimate the relative amount of yellow leaves visible from above by using equation (5.2) or (5.3), in which soil reflectance is replaced by the estimated reflectance of yellow vegetation.

During senescence the infrared reflectance will decrease (in a manner comparable with the influence of bare soil). This decrease in the infrared reflectance of a discolouring leaf will be more gradual: there will be no abrupt distinction between green or yellow leaves, as must be made when measuring LAI from harvested plants. Because the decrease in photosynthetic activity is also gradual, it is possible for the infrared reflectance to give a better estimate of the actual (photosynthetically active) LAI than field measurements on harvested plants. This is very difficult or nearly impossible to prove because one has to compare the reflectance measurements with the (subjective) field measurements.

If it is assumed that the ratios of reflectance values in different passbands for yellow vegetation can be estimated, it should be possible to use equation (5.25) to correct the infrared reflectance for the background of yellow leaves (the ratios now relate to yellow vegetation instead of to bare soil). When the infrared reflectance of yellow vegetation is at a similar level to that of either the green or the red reflectance of yellow vegetation or to both of these (compare situation with bare soil) it may be possible to ascertain the corrected infrared reflectance by using an equation analogous to (5.28). Finally, equation (5.12) may be used to estimate LAI. However, we now need to ascertain whether the two unknown parameters of this equation are the same as when the vegetation is green.

At the end of the season senescence may have advanced so far that the leaves shrivel and finally fall off. Then soil background will again be visible, providing a background with yellow and dead leaves. This will again affect reflectance in a manner that will be hard to describe. New experimental observations may open the way to estimating LAI from reflectance measurements.

## 5.8 Comparing the model with the SAIL model

In this section the model derivations presented earlier in this chapter will be verified and compared by means of calculations with the SAIL model (section 4.7). Special attention will be paid to verifying the various methods of correcting for soil background in estimating soil cover and LAI.

The following variables for the SAIL model have been used:

- three leaf angle distributions:
  spherical, planophile and erectophile.
- two irradiance conditions:
  direct sunlight only ($\theta_S = 45°$) and diffuse skylight only (in reality many intermediate situations will occur).
- three soil types:
  dry soil (green reflectance = 20.0 %, red reflectance = 22.0 %, infrared reflectance = 24.2 %);
  wet soil (green reflectance = 10.0 %, red reflectance = 11.0 %, infrared reflectance = 12.1 %);
  black soil (green, red and infrared reflectance = 0 %).
- reflectance and transmittance of a single leaf were assumed to be equal: green reflectance = 8 %, red reflectance = 4 % and infrared reflectance = 45 %.
- the direction of observation was assumed to be vertically downwards ($\theta_0 = 0°$).

Model calculations were carried out using the following LAI values: 0 (0.1) 1.0 (0.2) 2.0 (0.5) 5.0 (1.0) 8.0.

The green, red and infrared reflectance factors were calculated according to the SAIL model for each of the above situations. The model was also used to calculate soil cover with the conventional definition (vertical projection) and with the new definition introduced in section 5.2 for the situations with direct sunlight. The results of all model simulations are given in appendix 5. In order to limit the number of figures within this chapter, figures for the planophile and erectophile leaf angle distributions are mainly given in appendices.

### *5.8.1 Estimating soil cover*

The results obtained with the SAIL model (appendix 5) clearly show that the relationship between soil cover, according to the conventional definition, and green or red reflectance is non-linear (figure 5.10). However, using the new definition for soil cover, introduced in section 5.2, this relationship was nearly perfectly linear for all situations studied. The results for a dry soil and direct sunlight only are illustrated in figure 5.11 with a spherical leaf angle distribution, and in appendix 6 with a planophile and erectophile leaf angle distribution. Leaf angle distribution appeared to have only a minor influence on the reflectance at complete soil cover (see appendix 5), which was especially small for the red reflectance. Similar results were obtained for the situations with a wet soil.

These results support the validity of equations (5.2) and (5.3) and their usefulness for estimating soil cover if the new definition of soil cover that takes shadow and vegetation together is used.

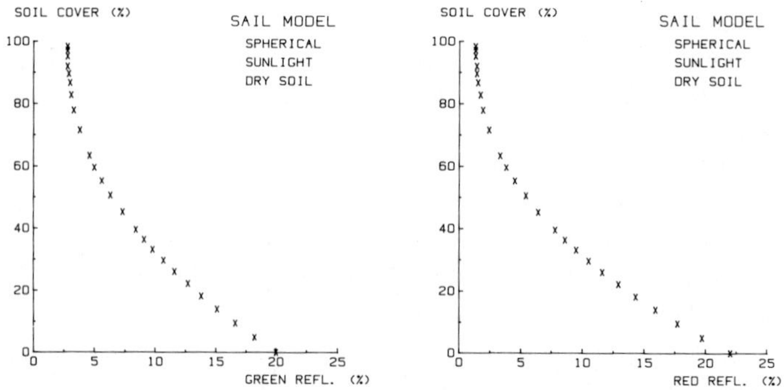


FIG. 5.10 An example of the relationship between soil cover, according to the conventional definition, and green or red reflectances, respectively.

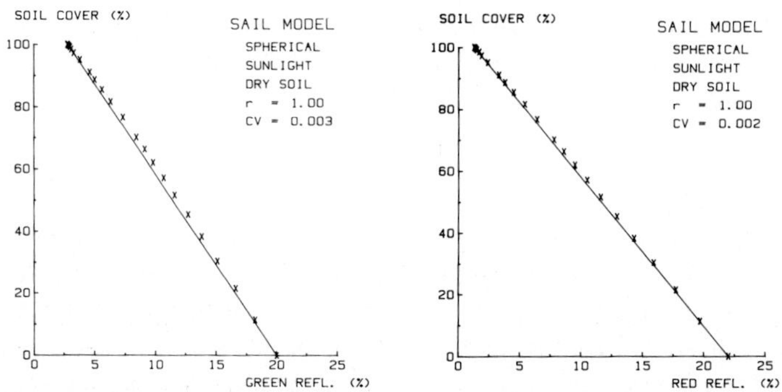


FIG. 5.11 Soil cover (new definition) as a function of green and red reflectance, respectively, for a spherical leaf angle distribution.
xx : calculated points SAIL model
— : simplified reflectance model

### *5.8.2 Correction for soil background in estimating soil cover*

In section 5.5 three methods were given for correcting the estimation of soil cover for differences in soil moisture content. Theoretically, method 1 should offer the best results (cf. section 5.5). This method (equation 5.20) takes the actual ratio of green and red reflectance of bare soil (which is assumed to be independent of soil moisture content) into account. Method 2 uses the difference between green and red reflectance (equation 5.21). Method 3 uses the ratio of green and red reflectance (equation 5.22). The latter two methods have only limited potential for correcting for differences in soil moisture content. The results for a spherical leaf angle distribution for dry and wet soil (direct sunlight only) are illustrated in figures 5.12 a and b, respectively, and with a planophile and an erectophile leaf angle distribution in appendix 7.

In all situations, method 3 gave the best fit to the data. However, up to about

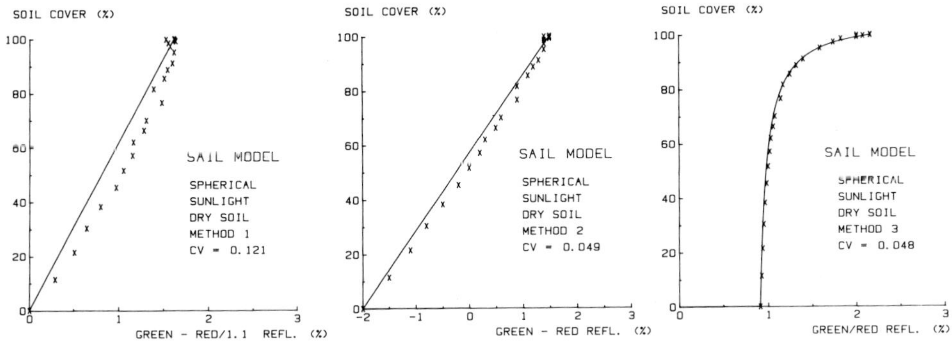


FIG. 5.12a Three methods for correcting for differences in soil moisture content in estimating soil cover. Spherical leaf angle distribution, dry soil.

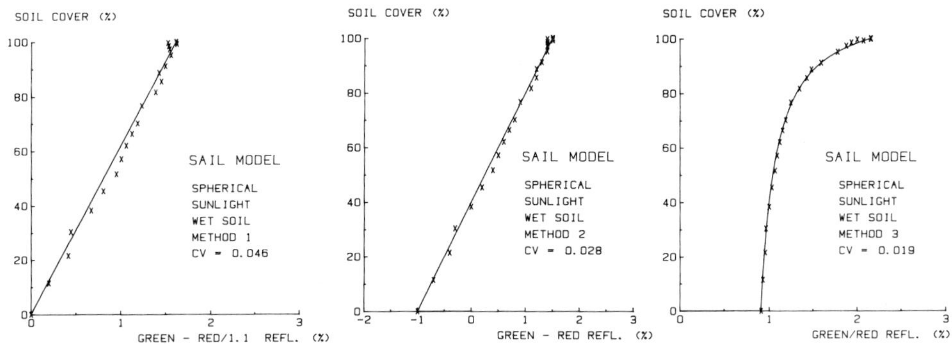


FIG. 5.12b Three methods for correcting for differences in soil moisture content in estimating soil cover. Spherical leaf angle distribution, wet soil.

80 % soil cover the ratio of green and red reflectance was nearly constant. This implies that a small inaccuracy in this ratio would induce a large inaccuracy in estimating soil cover. Accordingly, a small discrepancy between curves induced by a difference in soil moisture content, as illustrated in figure 5.13, would introduce a relatively large estimation error. Thus method 3 is less valuable for correction.

Method 2 gave a worse fit to the data than method 3. However, with method 2 there was a linear relationship between soil cover and the difference beteen green and red reflectance and it had a less steep slope than the slope at low soil cover with method 3. This implies that with method 2 the estimation of soil cover was less influenced by an inaccuracy in the difference between green and red reflectance than when the ratio between these two reflectances was used. However, for the reasons discussed in section 5.4 the difference itself was ascertained less accurately. Moreover, if the relationship between soil cover and reflectance is slightly non-linear (figure 5.11), this will severely influence the relationship between soil cover and the difference between reflectances. At low soil

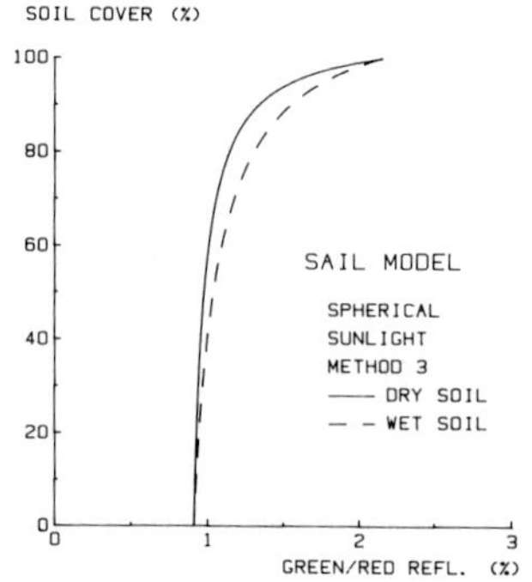


FIG. 5.13 Method 3 for correcting for differences in soil moisture content in estimating soil cover for a spherical leaf angle distribution.

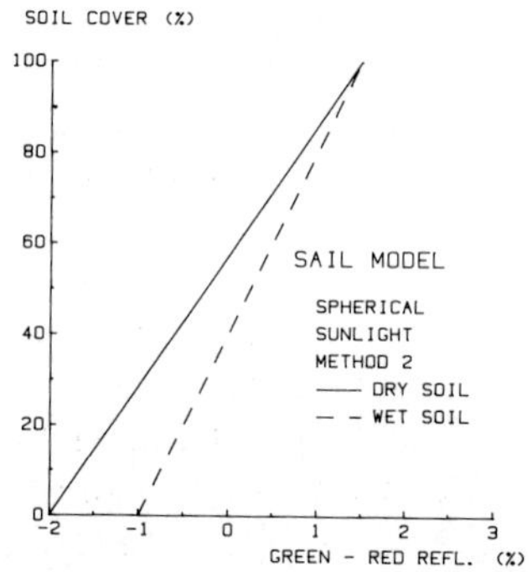


FIG. 5.14 Method 2 for correcting for differences in soil moisture content in estimating soil cover for a spherical leaf angle distribution.

cover values the influence of soil moisture content can be considerable (figure 5.14).

The fit with method 1 was the worst of all. Again, this is primarily because the difference between the reflectances was used. However, the position of the theoretically straight curve was independent of soil moisture content (figure 5.15). This may imply that the overall estimation error inherent in method 1 could be smaller than in methods 2 or 3, if one has to estimate soil cover without knowing soil moisture content or soil reflectance explicitly. When a mean soil reflectance of 15.0% in the green and 16.5% in the red was assumed, then in the case of the spherical leaf angle distribution thc coefficient of variation with method 3 was 0.071, with method 2 it was 0.093 and with method 1 it was 0.091. For the planophile leaf angle distribution the smallest coefficient of variation, namely 0.048, occurred with method 1. This coefficient was smaller than the values obtained with methods 2 or 3, which were 0.061 and 0.052, respectively.

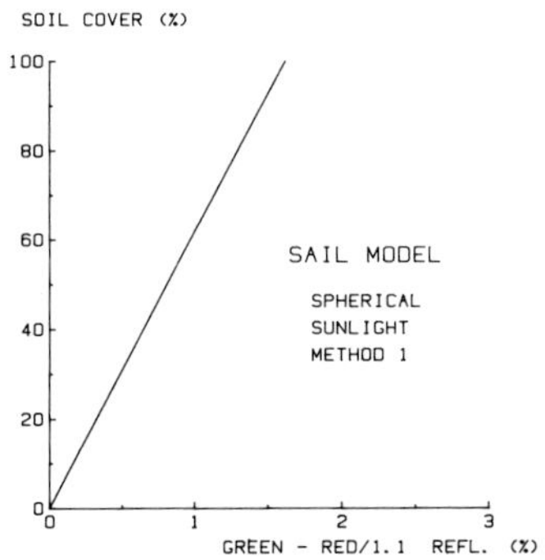


FIG. 5.15 Method 1 for correcting for differences in soil moisture content in estimating soil cover for a spherical leaf angle distribution.

### *5.8.3 Estimating LAI*

In estimating LAI the infrared reflectance is corrected for soil background and subsequently this corrected infrared reflectance is used for estimating LAI. This latter step may be investigated by using the calculations with the SAIL model for a black background. Then the infrared reflectance does not require correction and the validity of equation (5.12) may be checked. The results, shown in figure 5.16 and appendix 8, support the validity of this equation for describing the relationship between 'corrected' infrared reflectance and LAI at constant leaf angle distribution. The influences of the illumination conditions were only minor. It is notable that distinct leaf angle distributions cause quite distinct asymptotic values for the infrared reflectance, calculated from the SAIL model.

### *5.8.4 Correction for soil background in estimating LAI*

A correction can be made for differences in soil moisture content by subtract-

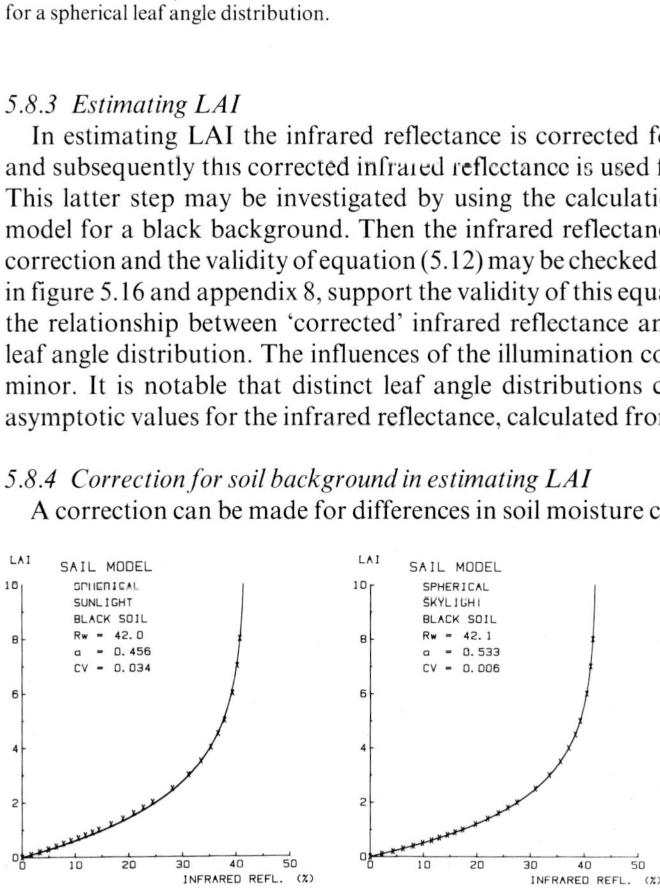


FIG. 5.16 LAI as a function of the infrared reflectance for a black soil with a spherical leaf angle distribution.

xx : calculated points SAIL model

— : simplified reflectance model

(Rw is used for $\hat{r}_{\infty,ir}$ and a is used for $\hat{\alpha}$ in these and subsequent graphs).

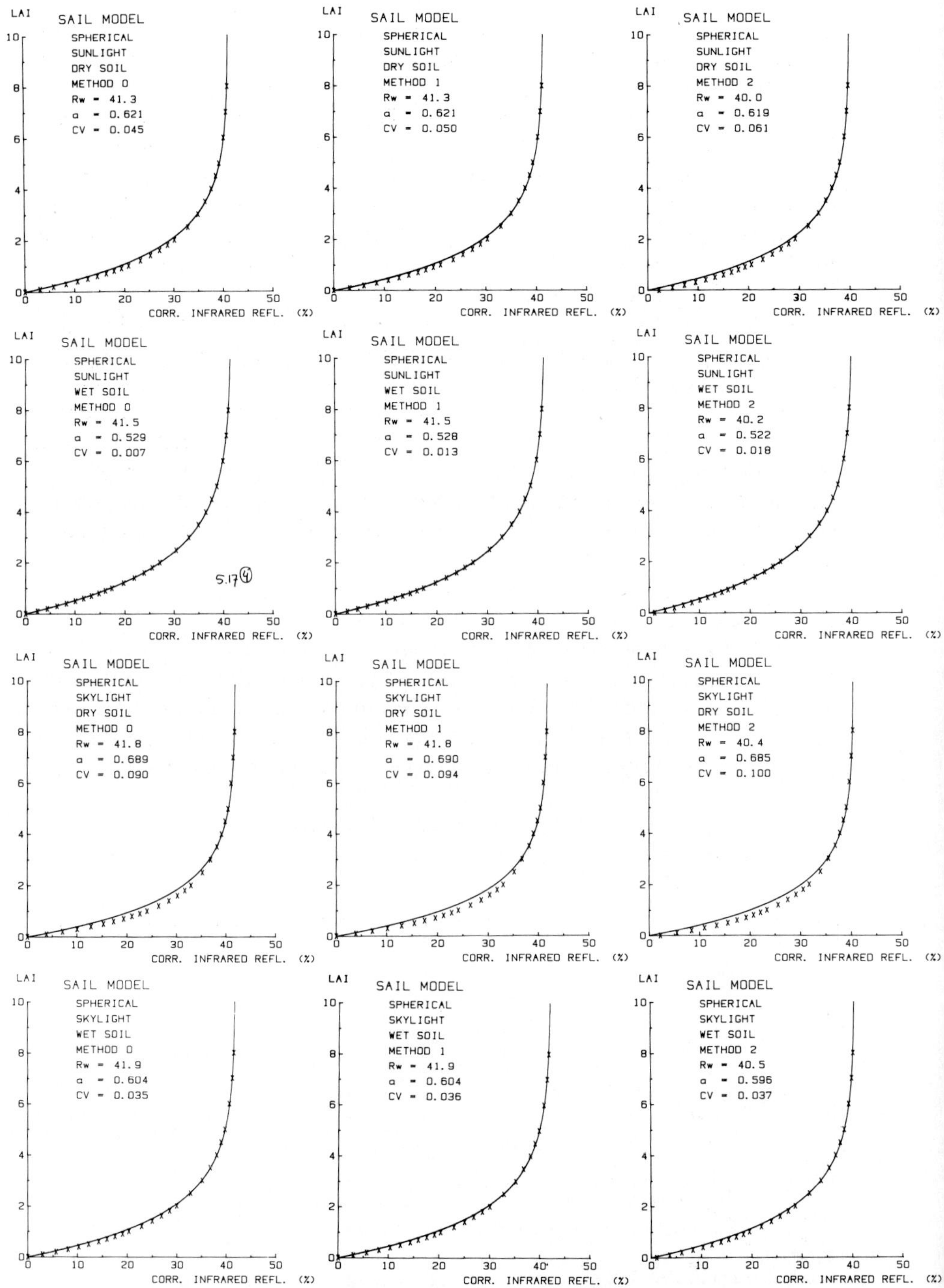


FIG. 5.17 Three methods for correcting for differences in soil moisture content in estimating LAI. Spherical leaf angle distribution.

ing the contribution of the soil visible to the eye from the measured infrared reflectance (equation 5.23). If soil reflectance is known, soil cover may be estimated by applying equation (5.16) e.g. at the red reflectance. This method of correcting the infrared reflectance will be called method 0 in this section (indicating that it cannot be applied without knowing soil reflectances explicitly). Method 1 takes into account the constant ratios of soil reflectance between passbands (equation 5.25). Method 2 ascertains the corrected infrared reflectance by taking the difference between infrared and red reflectance (equation 5.28) – a drastic simplification compared with method 1. Results for all three methods are given in figure 5.17 and appendices 9 and 10.

All three methods gave essentially the same results. The drastic simplification by method 2 yielded results that were, in general, not much worse than those obtained with the other methods. Because the only correction made is for soil visible to the eye and not for the soil underneath vegetation, some influence of soil background will still remain. This is illustrated in figure 5.18. Even with such a large range in soil reflectances, differences between curves were not very large. In reality, fluctuations in soil moisture content underneath vegetation will be less than those on bare soil. Finally, the small influence of the illumination conditions is illustrated in figure 5.19.

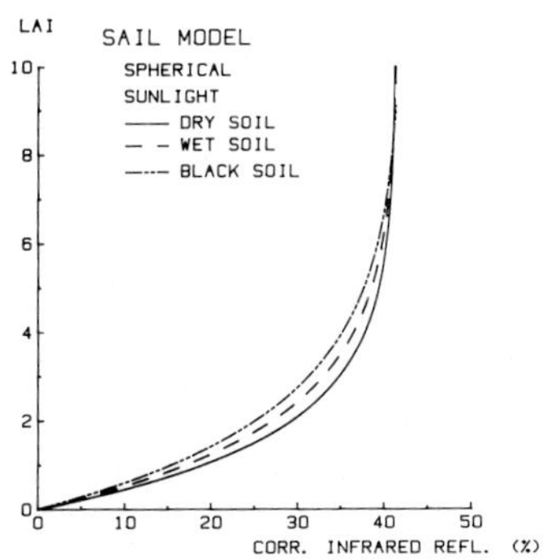


FIG. 5.18 Influence of soil background on the regression of LAI on corrected infrared reflectance.

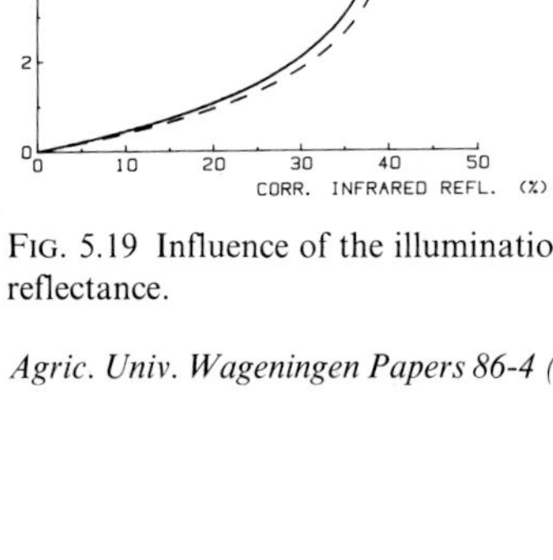


FIG. 5.19 Influence of the illumination conditions on the regression of LAI on corrected infrared reflectance.

## 5.9 Summary

In this chapter a new definition of soil cover has been introduced. Only soil that is illuminated by the sun as well as directly detectable by the sensor is classified as being not covered. Thus, if the sensor looks vertically downwards, soil cover is now defined as the relative vertical projection of green vegetation, the relative area of the shadows included.

A simplified reflectance model for vegetation has been introduced. It is based on equation 5.1:

$$r = r_V \cdot B + r_S \cdot (1-B)$$

Since $r_V$ is regarded to be a constant for visible passbands, the relationship between the measured reflectance in a visible passband and soil cover (new definition) is linear according to this equation. This was confirmed by model calculations with the SAIL model.

For estimating LAI the infrared reflectance will be corrected for the reflectance of the soil background. This corrected infrared reflectance has been defined as:

$$r'_{ir} = r_{ir} - r_{s,ir} \cdot (1-B)$$

Subsequently this corrected infrared reflectance is used for estimating LAI according to the following equation (equation 5.12):

$$L = -1/\alpha \cdot \ln(1 - r'_{ir}/r_{\infty,ir})$$

where $\alpha$ and $r_{\infty,ir}$ are 2 parameters that have to be estimated empirically from a training set. The applicability of this semi-empirical model for describing the relationship between LAI and corrected infrared reflectance was supported by model simulations with the SAIL model for a black background.

In order to estimate soil cover or LAI by using the above equations, the reflectance of bare soil must be known. This reflectance is to a large extent determined by soil moisture content. Often, the contribution of the soil to the measured reflectance is not explicitly known. In correcting for soil background three methods have been derived for estimating soil cover and two methods have been derived for estimating LAI. The main assumption has been that there is a constant ratio between the reflectances of bare soil in different passbands, independent of soil moisture content (equations 5.19 and 5.24). This assumption is valid for many soil types (cf. section 4.5), including the one used in this monograph (chapter 6).

For estimating soil cover an equation has been derived by combining reflectance measurements in a green and a red passband. This has been called method 1 (equation 5.20):

$$B = \frac{r_g - C_1 \cdot r_r}{r_{v,g} - C_1 \cdot r_{v,r}}$$

Here $C_1$ is the constant ratio between green and red reflectance factors of bare soil. Method 2 is a special case of method 1, as $C_1$ is assumed to be nearly equal to the value 1. With this method only the difference between green and red reflectance measurements has been used for estimating soil cover (equation 5.21):

$$B = \frac{(r_{s,g} - r_{s,r}) - (r_g - r_r)}{(r_{s,g} - r_{s,r}) - (r_{v,g} - r_{v,r})}$$

Both methods have the disadvantage that the difference between nearly equal reflectances in the green and red is used; this may result in the soil cover being grossly over- or underestimated (cf. section 5.4). Model calculations with the SAIL model indicated that a very small non-linearity of the relationship between reflectance and soil cover may lead to large discrepancies between the estimated soil cover with the SAIL model and the estimated soil cover using the above equations based on some difference function. A ratio function does not have this disadvantage (cf. section 5.4). Method 3 requires the ratio of green and red reflectance measurements for estimating soil cover (equation 5.22):

$$B = \frac{r_{s,g} - r_{s,r} \cdot r_g/r_r}{r_g/r_r \cdot (r_{v,r} - r_{s,r}) - (r_{v,g} - r_{s,g})}$$

However, method 3 is still influenced to some extent by reflectances of bare soil. Its (non-linear) regression curve has a steep slope at low soil cover, which is disadvantageous. Model calculations indicated that by assuming some mean reflectances for bare soil (with methods 2 and 3) all 3 methods yield comparable accuracies. Data from practice will have to show which method is best.

In order to estimate LAI, an equation correcting the infrared reflectance by combining reflectance measurements in green, red and infrared passbands has been derived. With method 1 the corrected infrared reflectance has been estimated as (equation 5.25):

$$r'_{ir} = r_{ir} - \frac{C_2 \cdot (r_g \cdot r_{v,r} - r_r \cdot r_{v,g})}{C_1 \cdot r_{v,r} - r_{v,g}}$$

Here $C_1$ and $C_2$ are the constant ratios between green and red reflectance and between infrared and red reflectance, respectively, of bare soil. Assuming that $C_1$ and $C_2$ are equal to 1, the corrected infrared reflectance can be approximated by (equation 5.28):

$$r'_{ir} = r_{ir} - r_r$$

This has been called method 2. Neither method requires soil reflectance as input. Model simulations with the SAIL model indicated that the accuracy obtained with both models corresponds to the one obtained if soil reflectances are known. Only the estimates of parameters $\alpha$ and $r_{\infty,ir}$ in estimating LAI differed slightly.

The main conclusion of this chapter is that for soil types that have a fairly uniform reflectance in the red and the adjacent infrared part of the electromagnetic spectrum, the difference between infrared and red reflectances provides a correction for soil background and can subsequently be used for estimating LAI. This latter relationship is described by an equation in which only 2 parameters have to be ascertained empirically.

The above model derivations are derived for a vegetative canopy. It is assumed that analogous derivations are valid for a generative canopy (cereals) with yellowing leaves (cf. results chapter 7). Then a correction for yellow leaves can be made. At the generative stage the relationship between corrected infrared reflectance and LAI will be described by estimates of parameters other than those used for the vegetative stage.

In chapter 7 the assumptions adopted in this chapter will be verified for one specific soil type and the different methods for estimating soil cover and LAI at agricultural field trials will be tested with real data. Moreover, it will be investigated whether the leaf angle distribution varied strongly in a multitemporal analysis, thus disturbing the regression of LAI on corrected infrared reflectance (cf. results SAIL model). But first, the field trials analysed and the methodology of data gathering and analysis will be described in chapter 6.

# 6 DATA GATHERING AND ANALYSIS

## 6.1 Introduction

This chapter begins by describing the field trials from which multispectral aerial photographic (MSP) recordings were obtained. The methods used to gather and analyse data will briefly be described. Finally, some general information about the missions flown will be given.

## 6.2 Agricultural field trials used in the present research

The field trials used to verify the usefulness of the remote sensing techniques detailed in preceding chapters were carried out on the ir. A.P. Minderhoudhoeve, experimental farm of the Wageningen Agricultural University (cf. section 1.4). These field trials are described below. Weeds were controlled in all trials. Some data from the meteorological station of the experimental farm are given in appendix 11.

### *6.2.1 Field trial 116 in 1982*

In 1982 field trial 116 was designed in order to investigate the influence of sowing date and nitrogen nutrition on the crop structure and the grain yield of barley. The trial was a split-plot design in three replicates with barley, cultivar 'Trumpf'. Whole-plot treatments were 2 sowing dates: 26 March (Z1) and 22 April (Z2). Split-plot treatments were 6 randomized nitrogen levels (applied before sowing):

N1 = 0 kg N per ha
N2 = 20 kg N per ha
N3 = 40 kg N per ha
N4 = 60 kg N per ha
N5 = 80 kg N per ha
N6 = 100 kg N per ha.

Each subplot was 6 m by 18 m and the row width was 13 cm.

On 12 dates during the growing season between May and the beginning of August, development stage and dry matter weight were ascertained at a frequency of approximately one week. Leaf area index (LAI) was measured only once fortnightly, resulting in six harvest dates. The sample size was 0.13 $m^2$. During flying missions the proportion of soil cover was estimated. On 17 August the barley was harvested by combine (45 $m^2$ per plot).

### *6.2.2 Field trial 116 in 1983*

In 1983 field trial 116 was designed in order to investigate the influence of sowing date and nitrogen nutrition on the plant development and the grain yield

A

B

C

FIG. 6.5 70-mm aerial photographs obtained at the ir. A. P. Minderhoudhoeve on 27 May 1982.
A: Green passband; PX 2402 with W21 + W57A filters
B: Red passband; PX 2402 with W70 filter
C: Infrared passband; IR 2424 with W87C filter.

Experimental design of field trials:

APM 116 BARLEY
Sowing dates:
Z1 = 26 March 1982
Z2 = 22 April 1982

Nitrogen levels (kg N per ha):
N1 = 0
N2 = 20
N3 = 40
N4 = 60
N5 = 80
N6 = 100

| | | |
|---|---|---|
| 1 Z1N5 | 13 Z2N1 | 25 Z1N3 |
| 2 Z1N2 | 14 Z2N2 | 26 Z1N4 |
| 3 Z1N3 | 15 Z2N5 | 27 Z1N1 |
| 4 Z1N6 | 16 Z2N6 | 28 Z1N6 |
| 5 Z1N4 | 17 Z2N4 | 29 Z1N5 |
| 6 Z1N1 | 18 Z2N3 | 30 Z1N2 |
| 7 Z2N2 | 19 Z1N4 | 31 Z2N6 |
| 8 Z2N6 | 20 Z1N2 | 32 Z2N1 |
| 9 Z2N5 | 21 Z1N3 | 33 Z2N3 |
| 10 Z2N3 | 22 Z1N1 | 34 Z2N5 |
| 11 Z2N1 | 23 Z1N6 | 35 Z2N4 |
| 12 Z2N4 | 24 Z1N5 | 36 Z2N2 |

APM 87 Rotation trial with POTATOES
Rotation
1 2 3 4

A = a – a    a = potatoes
B = a – z    z = spring wheat
C = a – s    s = sugar beet
D = a – z – s – h    h = oats

Nematicide
○ = untreated
+ = treated

| | | | | |
|---|---|---|---|---|
| 1 C2+ | 2 D4○ | 3 A2× | 4 A1+ | 5 B1○ |
| 6 A2○ | 7 B2○ | 8 D2○ | 9 C1+ | 10 D1+ |
| 11 D4+ | 12 C2○ | 13 D3○ | 14 D1○ | 15 D3+ |
| 16 B2+ | 17 D2+ | 18 C1○ | 19 B1+ | 20 A1○ |
| 21 A1○ | 22 C1○ | 23 B1+ | 24 D4+ | 25 A2+ |
| 26 D3+ | 27 D3○ | 28 D1○ | 29 C2+ | 30 B2+ |
| 31 B1○ | 32 C1+ | 33 D2+ | 34 D2○ | 35 D4○ |
| 36 D1+ | 37 A1+ | 38 B2○ | 39 A2○ | 40 C32○ |

of barley. The trial was a split-plot design in four replicates with barley, cultivar 'Trumpf'. Whole-plot treatments were 2 sowing dates: 7 March (Z1) and 21 April (Z2). Split-plot treatments were 5 randomized nitrogen levels, applied as split dressing (in kg N per ha):

| | Before sowing | Feekes stage 7 | Total |
|---|---|---|---|
| N1 | 0 | 0 | 0 |
| N2 | 20 | 0 | 20 |
| N3 | 20 | 20 | 40 |
| N4 | 20 | 40 | 60 |
| N5 | 20 | 60 | 80 |

Each subplot was 6 m by 20 m and the row width was 12.5 cm.

Development stage and dry matter weight were ascertained on 9 dates during the growing season and LAI was ascertained on 5 dates. The sample size was 0.125 $m^2$. Soil cover was estimated during flying missions. On 12 August the barley was harvested by combine (51 $m^2$ per plot).

*6.2.3 Field trial 100 in 1983*

Field trial 100 in 1983 was designed to investigate whether it is possible to use a few plots for ascertaining the regression of LAI on reflectance and then to use this regression curve for estimating LAI from reflectance measurements of the complete field trial. Some of the same treatments applied in field trial 116 were also applied in field trial 100. These treatments were N1 and N4 for both sowing dates (see section 6.2.2). The treatments were randomized within 2 complete replicates. Each subplot was 12 m by 20 m in order to allow larger samples to be harvested. Since it is easier to adjust days of field sampling to days of flying missions (the latter strongly depend on weather conditions), samples were obtained on 8 dates on which missions were flown. During the first two missions early in the season no samples were gathered because there was hardly any plant development. The sample size was 1.0 $m^2$.

*6.2.4 Field trial 92 in 1982*

In 1982 field trial 92 was designed in order to investigate the influence of plant density, nitrogen nutrition and fungicide treatment on development, grain filling and yield in spring wheat. The trial was a split-plot design in three replicates with spring wheat, cultivar 'Bastion' (sowing date: 26 March 1982). Whole-plot treatments were 2 fungicide treatments: no fungicides at all (F0) and 4 kg Bavistin M per ha at Feekes stage 5 combined with 0.5 l Bayleton per ha at stage 10.4 (F1). Split-plot treatments were 2 plant densities and 4 nitrogen levels, which were completely randomized within the whole plots. The sowing densities were: 150 seeds per $m^2$ (S1) and 300 seeds per $m^2$ (S2). The nitrogen levels were (in kg N per ha):

| | Before sowing | Feekes stage 8 | Total |
|---|---|---|---|
| N1 | 20 | 20 | 40 |
| N2 | 20 | 40 | 60 |
| N3 | 20 | 60 | 80 |
| N4 | 20 | 80 | 100 |

Each subplot was 6 m by 20 m and the row width was 13 cm.

Dry matter weight and LAI were ascertained on 7 dates during the growing season. The sample size was 0.13 m$^2$. On 18 August the spring wheat was harvested by combine (51 m$^2$ per plot).

### *6.2.5 Field trial 92 in 1983*

In 1983 field trial 92 was designed in order to investigate the influence of plant density, nitrogen nutrition and fungicide treatment on ear development and grain yield in winter wheat. The trial was a split-plot design in three replicates with winter wheat, cultivar 'Arminda' (sowing date: 25 October 1982). Whole-plot treatments were 2 fungicide treatments: no fungicides at all (F0) and 3 kg Bavistin M per ha at Feekes stage 5 combined with 2 kg Bayleton CF per ha at stage 10.4 (F1). Split-plot treatments were 3 plant densities and 4 nitrogen levels, which were completely randomized within the whole plots. The sowing densities were: 150 seeds per m$^2$ (S1), 300 seeds per m$^2$ (S2) and 600 seeds per m$^2$ (S3). The nitrogen levels were (in kg N per ha):

| | Feekes stage 8 | Total |
|---|---|---|
| N1 | 80 | 80 |
| N2 | 120 | 120 |
| N3 | 160 | 160 |
| N4 | 200 | 200 |

Each subplot was 6 m by 20 m and the row width was 12.5 cm.

Few data were obtained in the field. Dry matter weight and LAI were ascertained in the F1 plots on 12 April and 10 May, and in the S2 plots on 31 May, 28 June and 26 July. The sample size was 0.125 m$^2$. On 9 August winter wheat was harvested by combine (51 m$^2$ per plot).

### *6.2.6 Field trial 95 in 1983*

In 1983 field trial 95 was designed in order to investigate the influence of plant density, plant protection and nitrogen nutrition on tiller quality before flowering and on grain yield with winter wheat (cf. field trial 92 in 1982). The trial was a split-plot design in three replicates with winter wheat, cultivar 'Okapi' (sowing date: 25 October 1982). Whole-plot treatments were 2 fungicide treatments: no fungicides at all (F0) and 3 kg Bavistin M per ha at Feekes stage 5 combined with 2 kg Bayleton CF per ha at stage 10.4 (F1). Split-plot treatments were 3 plant densities and 4 nitrogen levels, which were completely randomized within the whole plots. The sowing densities were: 100 seeds per m$^2$ (S1), 200 seeds per m$^2$ (S2) and 400 seeds per m$^2$ (S3). The nitrogen levels were (in kg N per ha):

| | Feekes stage 3 | Feekes stage 8 | Total |
|---|---|---|---|
| N1 | 20 | 100 | 120 |
| N2 | 40 | 100 | 140 |
| N3 | 60 | 100 | 160 |
| N4 | 80 | 100 | 180 |

Each subplot was 6 m by 20 m and the row width was 12.5 cm.

Very few data on this trial were obtained in the field. LAI was ascertained in the F1 plots on 12 April and in the S2 plots on 31 May and 28 June. The sample size was 0.125 m$^2$. On 9 August winter wheat was harvested by combine (51 m$^2$ per plot).

*6.2.7 Field trial 85 in 1982*

Field trial 85 was designed in order to investigate the nitrogen utilization of crop systems with forage and arable crops. The trial was a split-plot design in 4 replicates. Whole-plot treatments were 9 different rotations with either grass, lucerne or maize. Split-plot treatments were 4 randomized nitrogen levels (in kg N per ha):

| | grass, lucerne | maize |
|---|---|---|
| N1 | 0 | 0 |
| N2 | 150 | 75 |
| N3 | 300 | 150 |
| N4 | 450 | 225 |

Each subplot was 12 m by 13 m.

On 18 May 1982 the rotations with grass were sampled (sample size 15 m$^2$). These were 3 rotations with grass lasting one (R1), two (R2) and three (R3) years, respectively. Fresh weight, dry matter weight and nitrogen yield were ascertained, as well as dry matter and nitrogen content.

*6.2.8 Field trial 87 in 1982*

Field trial 87 was designed in order to investigate the level and cause of damage to potatoes in various crop rotations. It was anticipated that yield would be considerably lower in narrow rotations with potatoes, because of nematodes. The trial was a completely randomized block design in two replicates. Potatoes, cultivar 'Hertha', were planted on 16 April 1982. Treatments were crop rotation and nematicide application (nematicide treatment (+) or none (o)). 4 different rotations were applied:
rotation A: potatoes only
rotation B: potatoes alternated with spring wheat
rotation C: potatoes alternated with sugar beet

rotation D: potatoes in a rotation with spring wheat, sugar beet and oats. Each plot was 6 m by 40 m.

Field measurements were very limited. On 21 June 1982, 6 plants per plot (potatoes) were harvested in order to ascertain the weight of foliage and tubers. On 20 September 1982 total yield (tubers) was ascertained.

## 6.3 Method of gathering data

The development stage of the crop in each field trial was recorded in the Feekes scale (appendix 1). Soil cover was estimated by taking vertical photographs from a height of 3-4 metres, overlaying the photographs with a grid and counting hits of soil per grid crossing. Dry matter weight and LAI were ascertained by harvesting all the plants within a 1.0 metre long section of row (0.13 $m^2$ or 0.125 $m^2$, depending on row width). After ascertaining fresh weight and tiller number of the whole sample, a subsample was separated into leaf blades (green and yellow), stems and ears. Each component was weighed, dried at 80 degrees Celsius and again weighed. Before drying, the area of the green leaf blades was measured with an optically scanning area meter. All the results were converted to give a value per square metre.

## 6.4 Data analysis

A very important analysis of field measurements from agricultural field trials is an analysis of variance in order to investigate whether treatment effects are significant and whether interactions occur. Analogously with field measurements, an analysis of variance can be carried out on reflectance measurements in the various passbands (chapter 3) for investigating whether the latter variables can be ascertained with relatively smaller variance and whether treatment effects can be ascertained with larger power than by means of conventional field measurements. Some concise remarks about the analysis of variance will be given in section 6.4.1.

The second main field of interest of this monograph is to investigate the possibilities of estimating crop characteristics by remote sensing. This particularly refers to the estimation of LAI by some corrected infrared reflectance factor (chapter 5).

One of the complications of sampling agricultural field trials is that this procedure is often destructive. To keep the plots fairly intact, only small samples are taken. The resulting variability of such data will be relatively large. These disturbances ('noise') may be decreased by data smoothing (section 6.4.2). In the present study, the means per treatment were smoothed. An important aspect of this smoothing is that it also involved an interpolation technique; thus, it was possible to estimate the LAI on every date during the growing season.

The smoothed estimated LAIs for the dates of flying missions were then related to reflectance measurements. Subsequently, the relationship between LAI (field measurements) and reflectance was used as a regression curve for estimating LAI per plot from the reflectance measurements. Finally, an analysis of variance was performed on these estimated LAI values. The results of this latter analysis are more comparable with the analysis of variance done on the original LAI measurements, since they involve the same variable.

### *6.4.1 Analysis of variance*

In field trials the aim is mostly to test whether there are treatment effects. As an example, consider a simple field trial with only one treatment (e.g. nitrogen nutrition). Any observed value (e.g. yield) is assumed to be the sum of three terms: (1) an overall mean, (2) a treatment deviation, and (3) a random element (residual effect) which is assumed to be a random sample from a normally distributed population. Accordingly, the analysis of variance partitions the sum of squares of the observations into three sums of squares, one attributable to the overall mean, one to the treatment effects, and one which is the residual sum of squares. In the analysis of variance the null hypothesis that no differences exist between the effects of the treatment is tested. This is done by calculating the ratio between the mean square of treatment effects and the residual mean square and subsequently testing this ratio, which is isomorous with an F statistic if the null hypothesis holds, with an F test at a certain level of significance (the 5% level is often used). In any statistical test the critical level (often called P-value) is often also given; this is the smallest level of significance at which the observed result would just lead to rejection of the null hypothesis (figure 6.1).

In general, the analysis of variance exerts two functions:

1. It is an elegant way of computing the pooled error variance ($s^2$) (= residual mean square). In comparing different procedures the coefficient of variation (cf. section 2.5.1) could be used.
2. It provides an F test of the null hypothesis that the population means are identical (absence of treatment effects) and indicates, by means of the critical level, the extent to which one should doubt the validity of the null hypothesis (the smaller the critical level, the stronger the evidence against the null hypothesis and therefore the more in favour of the presence of treatment effects).

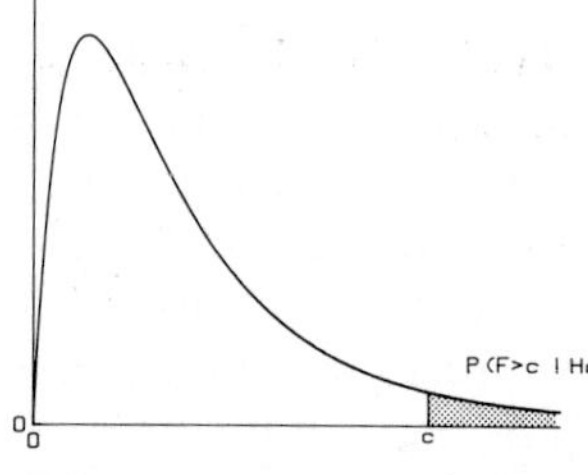


FIG. 6.1 Example of the distribution of the F-statistic. $P(F > c/Ho)$ is the critical level for $F = c$.

The probability that the null hypothesis will be rejected at a certain level of significance depends on the configuration of the true treatment effects and is called the power of the procedure. If of two procedures applied to the same data one procedure will lead to critical levels that are systematically smaller than those with the other procedure, then the former procedure has larger power.

The above example may be extended to more factors (e.g. plant density, sowing date), whereby interaction effects between factors can also be investigated (meaning that the effects of one factor are dependent on the levels of another factor, so that they are not merely additive).

A special experimental design must be used when it is necessary or convenient to test one factor (treatment) with experimental units of one size and to test a second factor with units of smaller size within the previous ones. This is called a split-plot design. The main factor may demand large areas for each level, e.g. irrigation. In an irrigation trial the trial area can be subdivided into main plots, some of which will be irrigated while others will not. These main plots may be subdivided into subplots for the application of other treatments, e.g. different levels of nitrogen nutrition, these levels being randomized within each main plot. The essential feature of such a design is that there are two (at least) types and sizes of experimental unit, resulting in two residual mean squares (corresponding to inter-plot and intra-plot variance): one attributable to whole-plot (or inter-plot) error and one to sub-plot (or intra-plot) error. The latter tends to be considerably smaller than the former. The main treatments are not compared as precisely as the subtreatments, for two reasons: (1) for main treatments less replication is provided and (2) subtreatment differences are not subject to whole-plot error. Hence, any subtreatment effect and the interactions with main treatments will be estimated more accurately. The split-plot design is especially suitable if one is mainly interested in the subfactor and the interaction, and less in the main factor.

For a more elaborate description of the analysis of variance, the reader is referred to books on statistics or experimental design (e.g. Cochran & Cox, 1950; Cox, 1958; Snedecor & Cochran, 1980).

### *6.4.2 Smoothing agronomic data*

Yield data can be smoothed by using a flexible family of growth curves. Because we observed an increasing standard deviation of disturbances with increasing growth of the crop, we adjusted growth curves to the logarithm of data, applying ordinary least squares.

A very flexible growth model is given by Schnute (1981). He describes the following solution for a general model with four parameters:

$$y = \left[ y_1^b + (y_2^b - y_1^b) \cdot \frac{1 - e^{-a \cdot (t - t_1)}}{1 - e^{-a \cdot (t_2 - t_1)}} \right]^{1/b} \tag{6.1}$$

with $a \neq 0$ and $b \neq 0$.

The four parameters are $y_1$, $y_2$, a and b. The variable y is the yield or size attained at time t; the parameter $y_1$ is the yield attained at time $t_1$ and $y_2$ that at time $t_2$. Time $t_1$ is designated to be somewhere at the beginning of the growing season, $t_2$ somewhere at the end. Schnute's family includes the Richards growth curves.

Schnute's growth model (equation 6.1) is based on the differential equation:

$$\frac{dr}{r \cdot dt} = -(a + b \cdot r) \tag{6.2}$$

$$\text{where } r = \frac{dy}{y \cdot dt} \tag{6.3}$$

Equation (6.3) defines the relative or specific growth rate. Equation (6.2) reflects the assumption that the relative growth rate of r is a linear function of r.

Because LAI may be regarded as being proportional to the derivative of the growth curve with respect to time (growth rate), the derivative of equation (6.1) may be used for smoothing LAI measurements.

Results of smoothing LAI data in this way indicated that parameter b was always small in the present study. Schnute has shown that for the limiting situation b=0 equation (6.1) specializes into:

$$y = y_1 \cdot e^{\left[\frac{1-e^{-a \cdot (t-t_1)}}{1-e^{-a \cdot (t_2-t_1)}} \cdot \ln\left(\frac{y_2}{y_1}\right)\right]} \tag{6.4}$$

If a>0 this submodel is equivalent to Gompertz growth (Schnute, 1981). The derivative of equation (6.4) with respect to time will be used for smoothing the LAI data according to the least squares method. These results are essentially the same as those with the four-parameter model (figure 6.2).

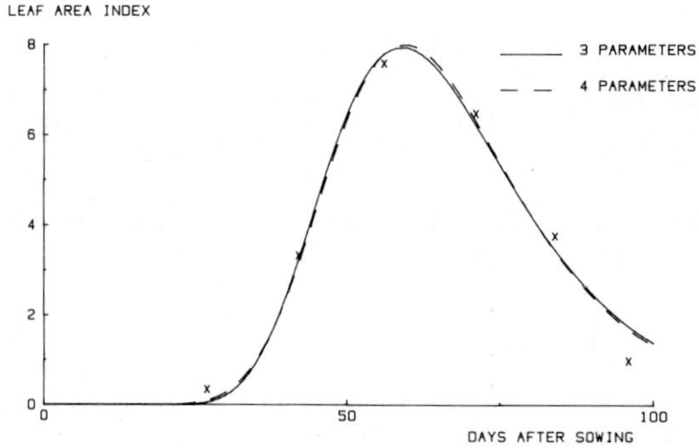


FIG. 6.2 Comparison of smoothing LAI data with a model with 3 and 4 parameters. Field trial 116 in 1982, treatment Z2N6.

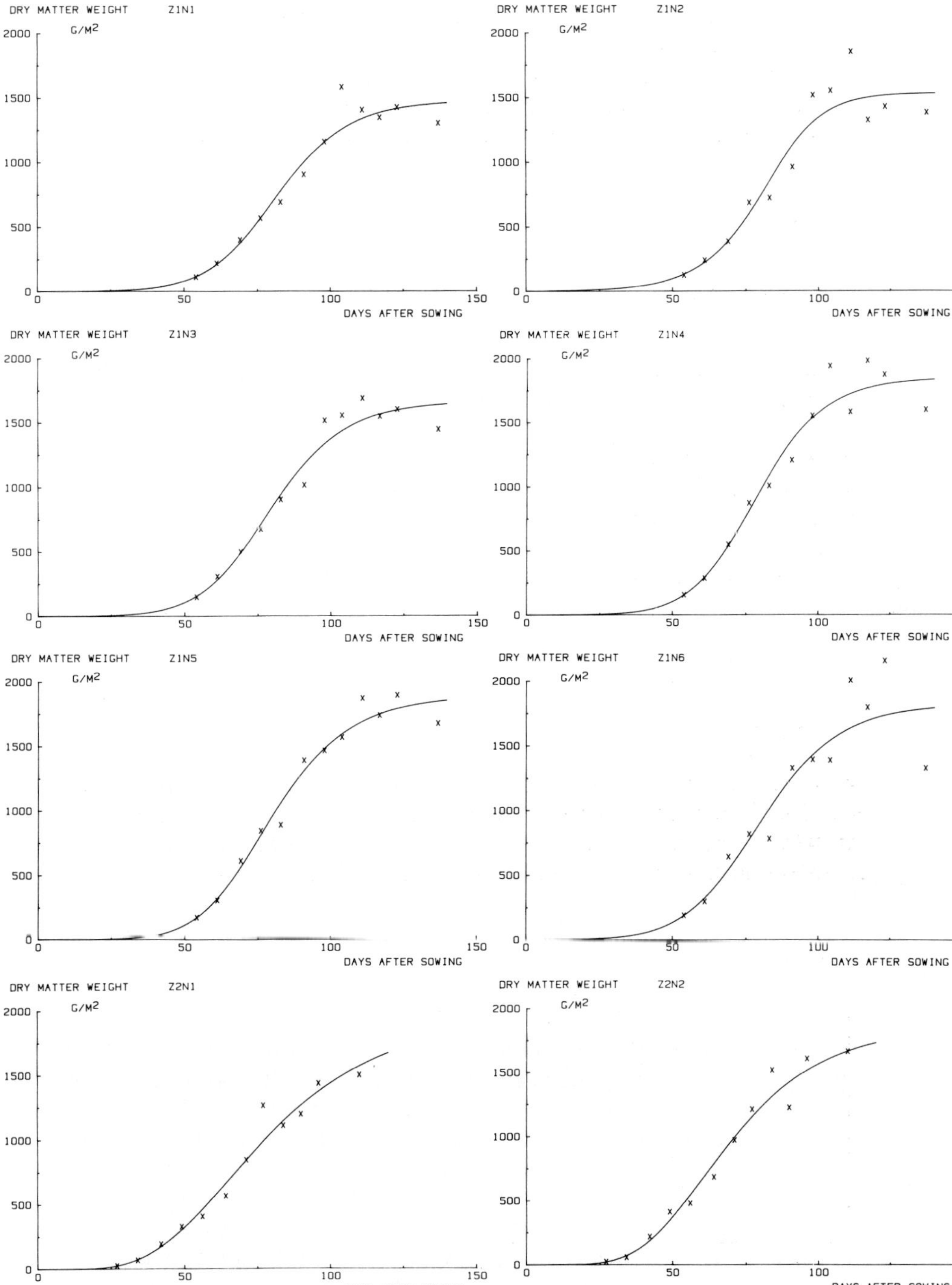


FIG. 6.3 Smoothing data on dry matter weight from field trial 116 in 1982.

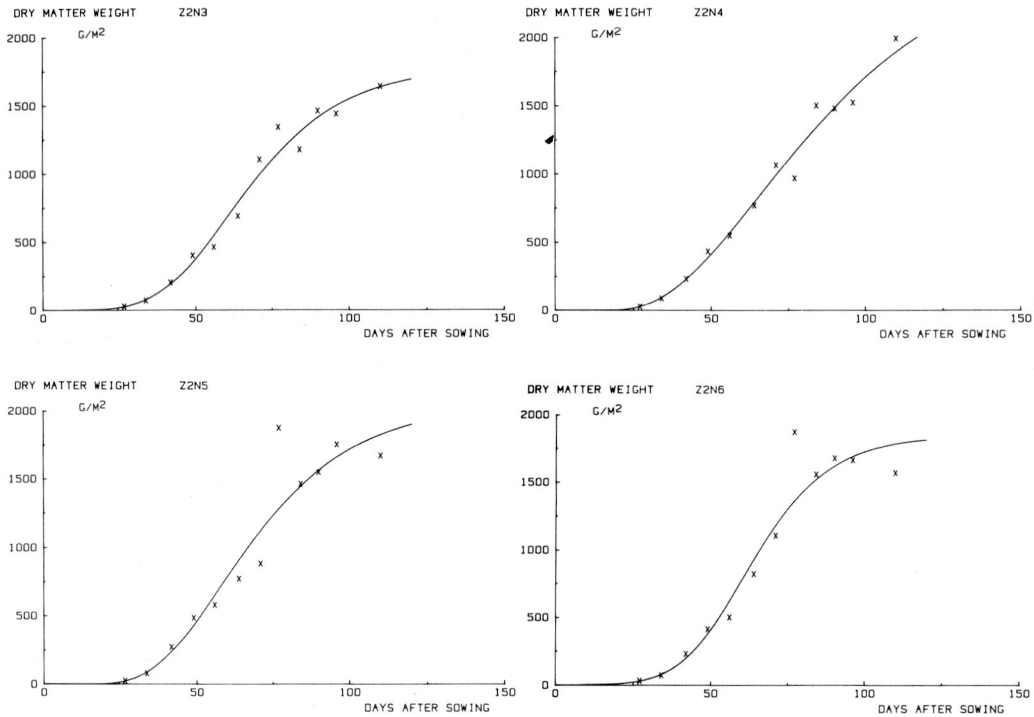


FIG. 6.3 continued.

Results of smoothing dry matter weight data according to equation (6.1) are illustrated in figure 6.3 for each of the 12 treatment combinations in field trial 116 in 1982, the least squares method being used for the logarithm of dry matter weight. Results of smoothing LAI data according to equation (6.4) are illustrated in figure 6.4 for the same 12 treatment combinations.

## 6.5 GENERAL INFORMATION ABOUT FLIGHTS

In this monograph, the results of multispectral aerial photography (MSP) during two growing seasons are analysed (chapter 7). The 1982 growing season being the first, some teething troubles were expected, for example because the crew had to become thoroughly acquainted with carrying out the missions, and also because of the problem of selecting the best weather conditions (cf. section 3.2). Also, exposing the film-material to the optimal density level offered some problems during this season, resulting in some overexposed recordings and thus to a loss in contrast (which did not, however, render the exposures useless). All these problems were solved before the 1983 growing season; thus, that season can be regarded as the first operational season.

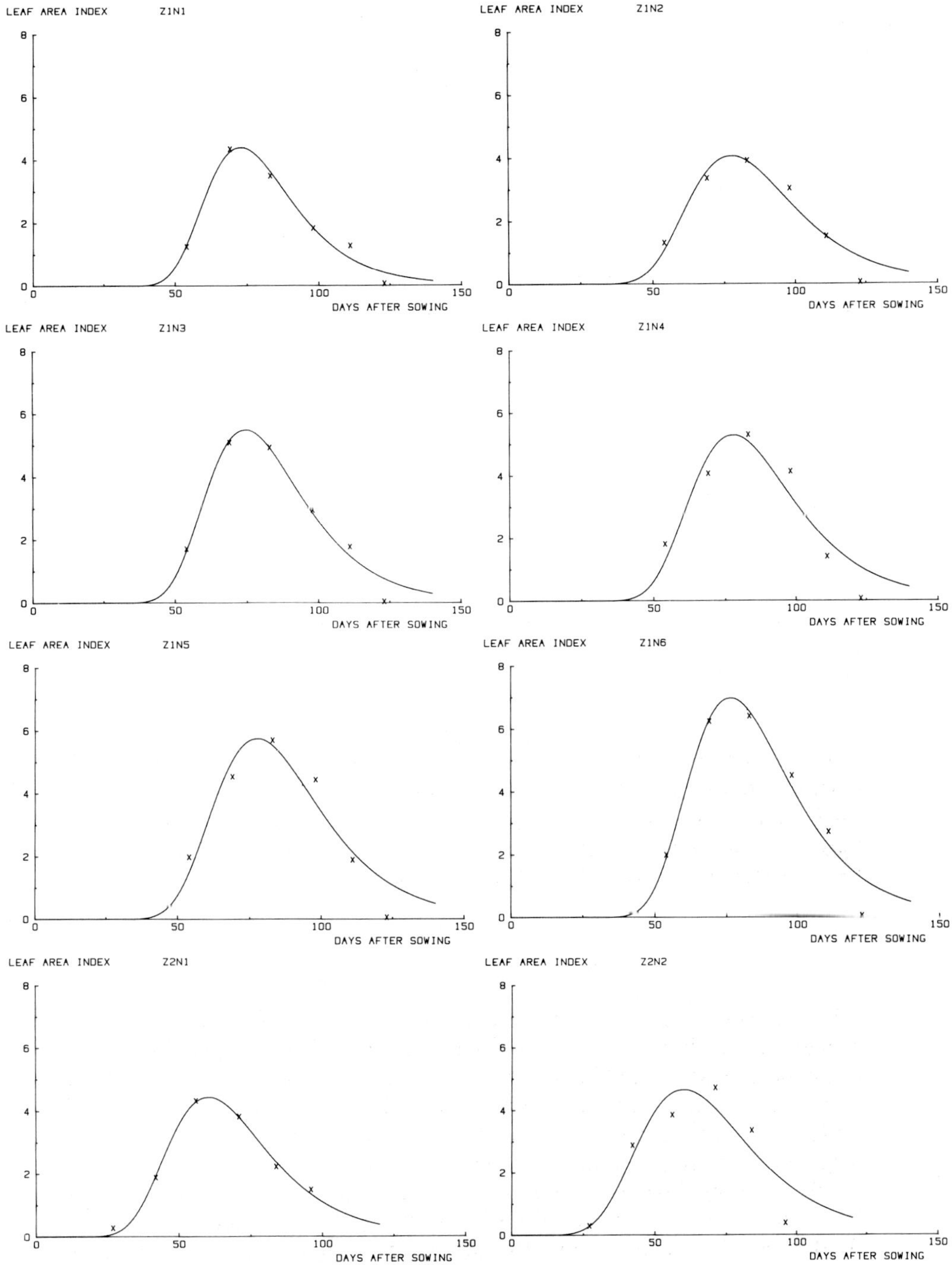


FIG. 6.4 Smoothing data on LAI from field trial 116 in 1982.

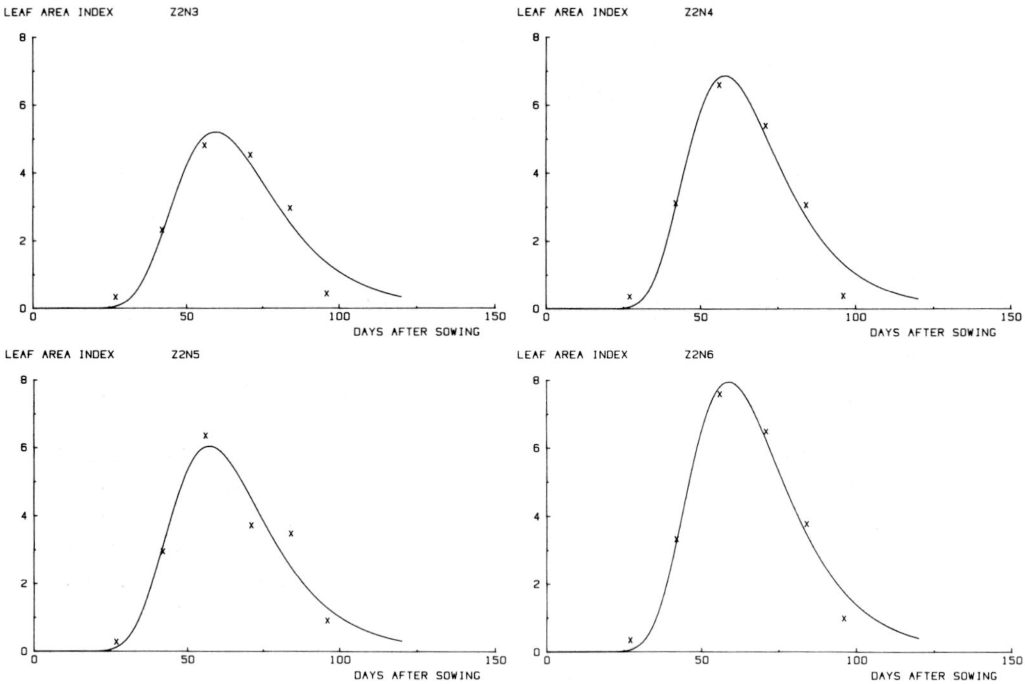


FIG. 6.4 continued.

Some information about the missions flown at the experimental farm during the 1982 and 1983 growing seasons is given in tables 6.1 and 6.2. On all dates exposures were obtained in the green, red and infrared passbands, except on 30 April 1983, when no exposures in the green passband could be obtained, because of the appearance of cumulus clouds. On 30 May 1983, illumination

TABLE 6.1 General information about the missions flown during the 1982 growing season.

| date | time (GMT)* | cloud conditions | lens type | remarks |
|---|---|---|---|---|
| 28 April | 09.30–10.30 | overcast sky | 50 mm | |
| 17 May | 09.00–10.00 | nearly cloudless | 50 mm | |
| 27 May | 09.00–10.00 | nearly cloudless | 100 mm | some cumulus clouds |
| 10 June | 14.00–15.00 | cloudless | 100 mm | very high cirrus |
| 17 June | 11.00–12.00 | overcast sky | 50 mm | |
| 1 July | 08.00–09.00 | overcast sky | 50 mm | |
| 13 July | 08.00–09.00 | cloudless | 100 mm | also MSS mission |
| 29 July | 08.30–09.30 | cloudless | 100 mm | |
| 11 August | 09.00–10.00 | cloudless | 100 mm | |

* GMT = Greenwich Mean Time

TABLE 6.2 General information about the missions flown during the 1983 growing season.

| date | time (GMT) | cloud conditions | lens type | remarks |
|---|---|---|---|---|
| 16 April | 08.30–09.30 | cloudless | 100 mm | |
| 30 April | 08.30–09.00 | cloudless | 100 mm | after 09.00 cumulus |
| 6 May | 08.30–09.30 | high cirrus | 100 mm | constant illumination |
| 30 May | 09.00–10.00 | overcast sky | 50 mm | slightly changing ill. |
| 7 June | 08.00–09.00 | cloudless | 100 mm | |
| 21 June | 08.00–09.00 | cloudless | 100 mm | |
| 5 July | 08.00–09.00 | cloudless | 100 mm | |
| 12 July | 08.00–09.00 | cloudless | 100 mm | |
| 22 July | 08.00–09.00 | cloudless | 100 mm | |
| 8 August | 10.30–11.30 | cloudless | 100 mm | |

could only be regarded as being constant for a couple of minutes. Since there was a time lag of about 10 minutes between taking exposures of field trials 92 and 95 and of the reference targets, this date was not incorporated into the analysis of both field trials. On that date the time lag between taking exposures of field trials 100 and 116 and the reference targets was only a few seconds. More meteorological information for both growing seasons is given in appendix 11 (days of missions are indicated by arrows). An example of multispectral aerial photography is offered in figure 6.5. (see page 106).

## 6.6 SUMMARY

In this chapter the design of eight agricultural field trials (1982 and 1983 growing seasons) has been described. In chapter 7 these trials will be used for analysing field measurements and spectral reflectance measurements and for ascertaining their relationship.

For comparing the conventional variables measured in the field and the variables measured by remote sensing techniques, coefficients of variation may be used as a measure of inaccuracy, and critical levels may be given for indicating the power with which treatment effects can be ascertained. Both result from analyses of variance.

For relating reflectance factors to LAI, the LAI data may be smoothed by using a growth model given by Schnute (1981). This smoothing reduces the variability within the LAI data and it allows the LAI to be estimated for the dates of flying missions from field measurements on intermediate dates.

Both the analysis of variance and the smoothing method will be applied to the results, as presented in chapter 7.

# 7 RESULTS

## 7.1 Introduction

In this chapter the results obtained from the multispectral aerial photography (MSP) described in chapter 3, are presented for several field trials. The results obtained for field trials with barley (field trial 116 in 1982 and field trials 100 and 116 in 1983) are described in detail. For field trials with wheat (field trial 92 in 1982 and field trials 92 and 95 in 1983) only the main results are described in detail and general results are presented in appendices, with references to these appendices in the main text.

Firstly, some attention is paid to the reflectance of bare soil (section 7.2) in order to verify the main assumption (as stated in chapter 5) that the ratio between the reflectances of bare soil in two passbands is constant.

In section 7.3 results are presented in terms of soil cover and LAI, as ascertained by field measurements. In section 7.4 results are presented of reflectance factors in the green, red and infrared passbands. Interrelationships are described. Also, measures of inaccuracy (in terms of coefficients of variation) and critical levels in testing the effects of treatments are compared by carrying out analyses of variance on the various variables.

In section 7.5 the possibilities for estimating soil cover (as defined in section 5.2) with aerial photography are investigated by using the methods derived in section 5.5.

In section 7.6 the possibilities for estimating LAI (the main crop characteristic) with aerial photography are investigated by using the methods derived in section 5.6. A regression curve is calculated by using smoothed LAI data and reflectance data for dates of missions. Using this regression curve the LAI per plot is estimated from reflectance values. By applying an analysis of variance to the latter LAI values obtained from remote sensing, the power (cf. section 6.4.1) of these estimations of expected LAI is compared with that obtained from field measurements of LAI. Finally, the possibilities of using only a limited number of plots (a training set) for ascertaining the regression curve of LAI on reflectance, and subsequently estimating LAI for an entire field trial, are also investigated.

Field trials 85 and 87 are described separately at the end of this chapter (sections 7.7 and 7.8, respectively).

Field measurements for field trial 85 were obtained by research workers from the Centre for Agrobiological Research in Wageningen. Field measurements for all other field trials used were obtained by research workers from the Department of Field Crops and Grassland Science of the Agricultural University. MSP recordings were made by personnel of the Department of Landsurveying and Remote Sensing of the Agricultural University. Personnel from the Ir. A.P. Minderhoudhoeve assisted in all the field trials.

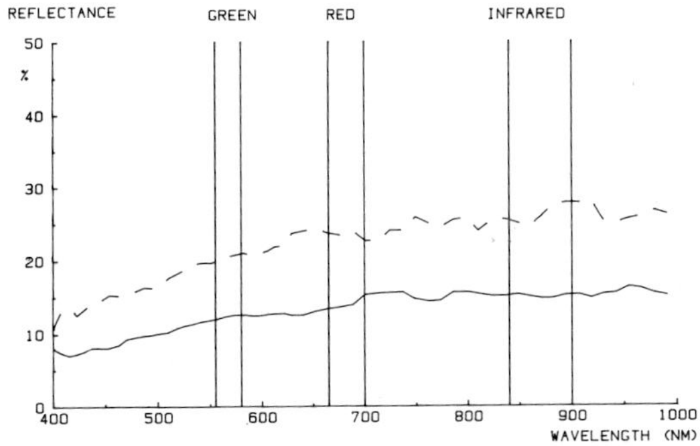


FIG. 7.1 Reflectance of bare soil as measured by means of the field spectroradiometer on two dates during the 1982 growing season.
—:wet soil (28 April 1982)
---:dry soil (10 June 1982)

## 7.2 REFLECTANCE OF BARE SOIL

The main assumption introduced in chapter 5 is that there is a constant ratio between the reflectance factors of bare soil in two passbands, which is independent of soil moisture content for the soil type used in this research.

During the 1982 growing season, reflectance measurements of bare soil were obtained with the field spectroradiometer. Results are illustrated in figure 7.1 for a wet soil and a dry soil. Estimated reflectances for the (simulated) passbands used with MSP are listed in table 7.1.

Some results of reflectance measurements of bare soil obtained by means of the hand-held radiometer during the 1983 growing season are listed in table 7.2. Spring 1983 was too wet for carrying out measurements with the field spectroradiometer.

The results obtained from both the field spectroradiometer and the hand-held radiometer confirm the validity of the assumption that the ratio between the reflectance in any pair of the green, red and infrared passbands is constant.

TABLE 7.1 Estimated reflectances of a wet soil and a dry soil for the MSP passbands, ascertained by means of the field spectroradiometer.

| | reflectance (%) | | | | |
|---|---|---|---|---|---|
| | green | red | infrared | green/red | infrared/red |
| 28 April 1982 | 12.4 | 13.6 | 15.0 | 0.91 | 1.10 |
| 10 June 1982 | 20.5 | 23.7 | 26.3 | 0.86 | 1.11 |

TABLE 7.2 Reflectance measurements of bare soil obtained with the hand-held radiometer (see section 2.4 for a description of the passbands).

| | reflectance (%) | | | | |
|---|---|---|---|---|---|
| | green | red | infrared | green/red | infrared/red |
| 30 March 1983 | 7.0 | 8.3 | 9.2 | 0.84 | 1.11 |
| 5 April 1983 | 16.5 | 17.4 | 19.3 | 0.95 | 1.11 |
| 15 April 1983 | 11.1 | 12.3 | 14.9 | 0.90 | 1.21 |
| 19 April 1983 | 14.3 | 15.9 | 17.6 | 0.90 | 1.11 |
| 22 April 1983 | 7.3 | 8.3 | 8.8 | 0.88 | 1.06 |
| 30 April 1983 | 9.0 | 10.3 | 12.1 | 0.87 | 1.17 |
| 6 May 1983 | 20.8 | 22.0 | 23.2 | 0.95 | 1.05 |

## 7.3 GROWTH VARIABLES

### 7.3.1 *Soil cover*

During missions, soil cover (new definition as described in section 5.2) was ascertained at nitrogen levels N1, N3 and N6 of field trial 116 in 1982 and in all treatments of field trials 116 and 100 in 1983. Estimates of soil cover were made in one replicate per trial until complete soil cover was reached. In the other field trials soil cover was not ascertained (see section 6.2). The results of these estimates for field trials 116 in 1982, 116 in 1983 and 100 in 1983 are given in tables 7.3, 7.4 and 7.5, respectively.

TABLE 7.3 Estimates of percentage of soil cover for field trial 116 in 1982 with barley, cultivar Trumpf.

| date | Z1N1 | Z1N3 | Z1N6 | Z2N1 | Z2N3 | Z2N6 |
|---|---|---|---|---|---|---|
| 17 May | 70 | 80 | 90 | 20 | 20 | 25 |
| 27 May | 90 | 100 | 100 | 50 | 40 | 50 |
| 10 June | | complete cover | | 85 | 95 | 100 |
| 17 June | | | | | complete cover | |

TABLE 7.4 Estimates of soil cover (%) for field trial 116 in 1983 with barley, cultivar Trumpf.

| date | Z1 | | | | | Z2 | | | | |
|---|---|---|---|---|---|---|---|---|---|---|
| | N1 | N2 | N3 | N4 | N5 | N1 | N2 | N3 | N4 | N5 |
| 14 April | 5 | 5 | 5 | 5 | 5 | 0 | 0 | 0 | 0 | 0 |
| 30 April | 20 | 25 | 25 | 25 | 25 | 0 | 0 | 0 | 0 | 0 |
| 6 May | 25 | 30 | 30 | 30 | 35 | 2 | 2 | 2 | 2 | 2 |
| 30 May | 75 | 80 | 90 | 90 | 90 | 40 | 40 | 40 | 40 | 40 |
| 7 June | 90 | 95 | 100 | 100 | 100 | 60 | 70 | 70 | 75 | 70 |
| 21 June | | | complete cover | | | 95 | 90 | 95 | 95 | 80 |
| 5 July | | | | | | | | complete cover | | |

TABLE 7.5 Estimates of soil cover (%) for field trial 100 in 1983 with barley, cultivar Trumpf.

| date | Z1N1 | Z1N4 | Z2N1 | Z2N4 |
|---|---|---|---|---|
| 14 April | 5 | 10 | 0 | 0 |
| 30 April | 25 | 30 | 0 | 0 |
| 6 May | 35 | 50 | 2 | 2 |
| 30 May | 70 | 85 | 40 | 40 |
| 7 June | 85 | 90 | 60 | 70 |
| 21 June | complete cover | | 90 | 95 |
| 5 July | | | complete cover | |

### 7.3.2 *Leaf area index*

Leaf area index (LAI) was only ascertained on 6 dates in the growing season in field trial 116 in 1982. Figure 7.2 presents the effects of the different treatments on each date. Only at the beginning of the season did the LAI of the early-sown crop (Z1) exceed that of the late-sown crop (Z2). In the middle of the season the late-sown crop made rapid progress and overtook the early-sown crop in July. At that stage the early-sown crop had already started to yellow. A positive nitrogen effect was most pronounced for the early-sown crop. Figure 7.3 pre-

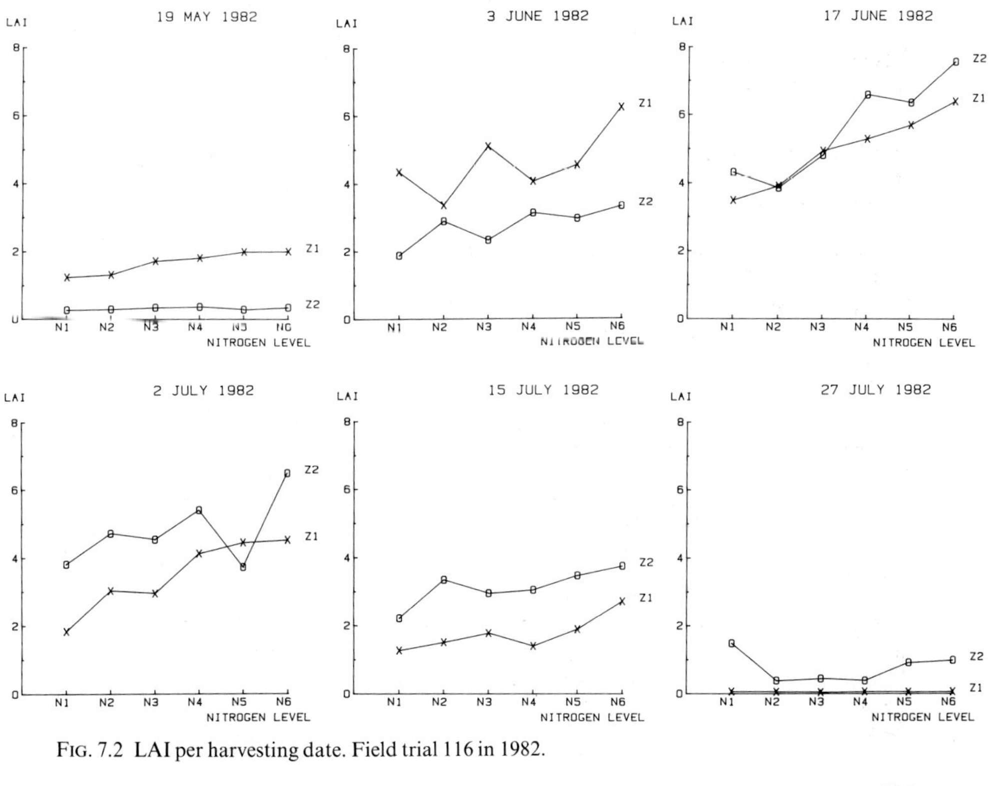


FIG. 7.2 LAI per harvesting date. Field trial 116 in 1982.

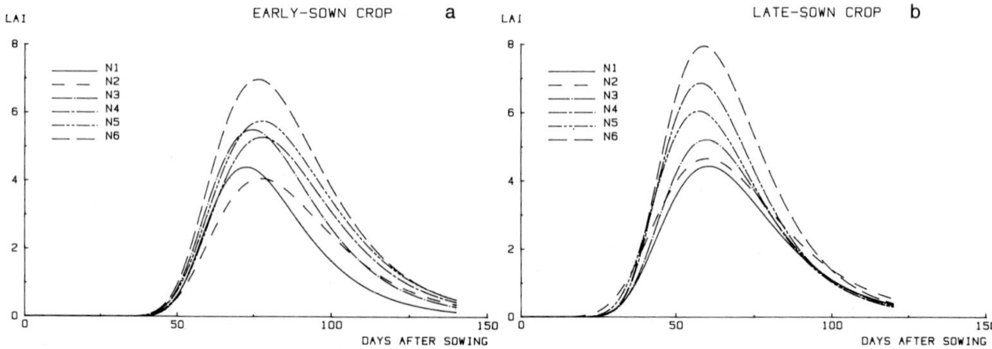


FIG. 7.3 Smoothed curves of LAI for the early-sown (a) and the late-sown (b) crop. Field trial 116 in 1982.

sents the smoothed LAI per sowing date. Intersections between these curves, especially in the early-sown crop, may be the result of inaccurate sampling. Nevertheless, the beneficial effect of increasingly large doses of nitrogen is clear for both crops, although in the late-sown crop N4 and N5 changed places. This may again be the result of inaccurate sampling.

In field trial 116 in 1983 LAI was ascertained on 5 dates in the growing season. Figure 7.4 presents the effects of the different treatments per date. At the beginning of the season the LAI of the early-sown crop was larger than the LAI of the late-sown crop, as expected. At the beginning of June the late-sown crop caught up with the early-sown crop, resulting in the larger LAI for the late-sown crop at the end of the season. Figure 7.4 shows clearly the effect of nitrogen nutrition before sowing on LAI (N1 as apposed to N2, N3, N4 and N5). During the second half of the season there is some effect of nitrogen given later in the season (at F7). Figure 7.5 presents the smoothed LAI data for each sowing date. It is noteworthy how many intersections occur, even after smoothing.

Results of LAI measurements of spring wheat in field trial 92 in 1982 and of winter wheat in field trials 92 and 95 in 1983 are given in appendices 12, 13 and 14, respectively.

Table 7.6 lists the means ($\bar{x}$), estimated coefficients of variation (CV) of residuals ($s/\bar{x}$) and the critical levels in testing for treatment effects, for the LAI in field trial 116 in 1982. The variance was determined by means of appropriate pooling of inter-plot and intra-plot variances. On 19 May the interaction effect between sowing date and nitrogen nutrition was significant at the 5% level. The sowing date effect was significant at the end of the season (from 2 July onwards), and the nitrogen nutrition effect was significant during the middle of the growing season (17 June up to 15 July). Coefficients of variation were quite large, especially the one on the last sampling date (27 July). This was caused by the

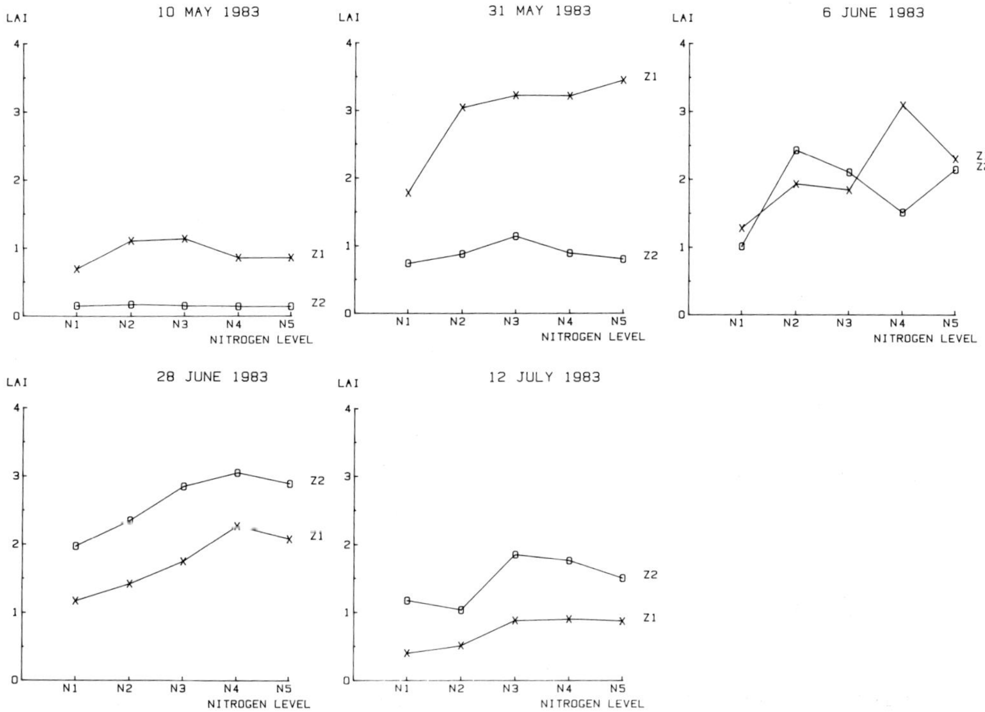


FIG. 7.4 LAI per harvesting date. Field trial 116 in 1983.

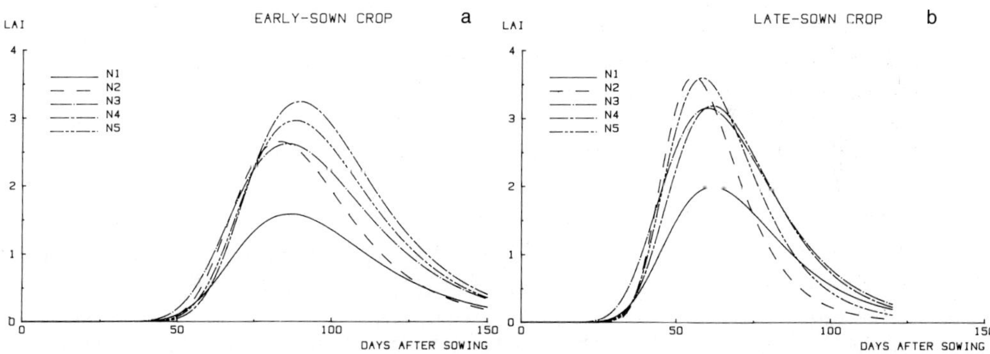


FIG. 7.5 Smoothed curves of LAI for the early-sown (a) and the late-sown (b) crop. Field trial 116 in 1983.

measurement of LAI becoming increasingly inaccurate as senescence proceeded (as stated in section 1.1).

Table 7.7 lists the means, the estimated CVs and the critical levels in testing for treatment effects, for the LAI in field trial 116 in 1983. Again, CV values for LAI were quite large.

TABLE 7.6 LAI: means, CVs and critical levels in testing for treatment effects, obtained on consecutive sampling dates. Field trial 116 in 1982 (36 plots).

| date | mean | CV | critical level in testing: | | |
|---|---|---|---|---|---|
| | | | inter-action | sowing dates | nitrogen nutrition |
| 19 May | 0.99 | 0.215 | 0.030 | 0.003 | 0.011 |
| 3 June | 3.67 | 0.363 | 0.474 | 0.092 | 0.277 |
| 17 June | 5.27 | 0.256 | 0.920 | 0.123 | 0.004 |
| 2 July | 4.13 | 0.221 | 0.150 | 0.044 | 0.002 |
| 15 July | 2.45 | 0.248 | 0.738 | 0.020 | 0.012 |
| 27 July | 0.41 | 1.301 | 0.457 | 0.017 | 0.445 |

TABLE 7.7 LAI: means, CVs and critical levels in testing for treatment effects, obtained on consecutive sampling dates. Field trial 116 in 1983 (40 plots).

| date | mean | CV | critical level in testing: | | |
|---|---|---|---|---|---|
| | | | inter-action | sowing dates | nitrogen nutrition |
| 10 May | 0.54 | 0.420 | 0.283 | 0.003 | 0.208 |
| 31 May | 1.91 | 0.353 | 0.010 | 0.030 | 0.001 |
| 6 June | 1.96 | 0.288 | 0.008 | 0.356 | 0.002 |
| 28 June | 2.18 | 0.272 | 0.980 | 0.018 | 0.007 |
| 12 July | 1.09 | 0.338 | 0.788 | 0.001 | 0.007 |

TABLE 7.8 LAI: means, CVs and critical levels in testing for treatment effects, obtained on consecutive sampling dates. Field trial 92 in 1982 (48 plots).
A = fungicide treatment
B = sowing density
C = nitrogen nutrition
ABC, AB, AC, BC = interactions

| date | mean | CV | critical level in testing: | | | | | | |
|---|---|---|---|---|---|---|---|---|---|
| | | | ABC | AB | AC | BC | A | B | C |
| 18 May | 1.23 | 0.209 | 0.057 | 0.837 | 0.122 | 0.892 | 0.473 | 0.000 | 0.644 |
| 1 June | 2.69 | 0.279 | 0.615 | 0.984 | 0.564 | 0.162 | 0.301 | 0.014 | 0.787 |
| 15 June | 3.63 | 0.212 | 0.152 | 0.630 | 0.114 | 0.151 | 0.204 | 0.004 | 0.278 |
| 30 June | 3.78 | 0.204 | 0.132 | 0.950 | 0.023 | 0.131 | 0.071 | 0.001 | 0.576 |
| 13 July | 3.01 | 0.267 | 0.568 | 0.617 | 0.405 | 0.080 | 0.087 | 0.083 | 0.473 |
| 27 July | 0.44 | 0.530 | 0.266 | 0.421 | 0.599 | 0.310 | 0.024 | 0.442 | 0.028 |
| 10 August | 0.00 | – | – | – | – | – | – | – | – |

TABLE 7.9 LAI: means, CVs and critical levels in testing for treatment effects, obtained on consecutive sampling dates. Field trial 92 in 1983 (total of 72 plots). (For explanation of A, B, C etc. see table 7.8)

| date | treat-ments harvested | mean | CV | critical level in testing: | | | | | |
|---|---|---|---|---|---|---|---|---|---|
| | | | | AB | AC | BC | A | B | C |
| 12 April | F1 | 0.69 | 0.263 | – | – | 0.052 | – | 0.000 | 0.564 |
| 10 May | F1 | 1.68 | 0.230 | – | – | 0.381 | – | 0.133 | 0.189 |
| 31 May | S2 | 2.28 | 0.266 | – | 0.763 | – | 0.663 | – | 0.247 |
| 28 June | S2 | 2.79 | 0.148 | – | 0.100 | – | 0.415 | – | 0.086 |
| 26 July | S2 | 0.19 | 0.794 | – | 0.168 | – | 0.010 | – | 0.168 |

TABLE 7.10 LAI: means, CVs and critical levels in testing for treatment effects, obtained on consecutive sampling dates. Field trial 95 in 1983 (total of 72 plots). (For explanation of A, B, C etc. see table 7.8)

| date | treat-ments harvested | mean | CV | critical level in testing: | | | | | |
|---|---|---|---|---|---|---|---|---|---|
| | | | | AB | AC | BC | A | B | C |
| 12 April | F1 | 0.69 | 0.240 | – | – | 0.959 | – | 0.000 | 0.559 |
| 31 May | S2 | 2.98 | 0.176 | – | 0.999 | – | 0.857 | – | 0.764 |
| 28 June | S2 | 2.31 | 0.260 | – | 0.740 | – | 0.020 | – | 0.618 |

Results of the analysis of variance for LAI measurements in field trial 92 in 1982 are listed in table 7.8. Few LAI measurements were gathered in field trials 92 and 95 in 1983 (see section 6.2.5 and 6.2.6). The results of the analysis of variance for the LAI data obtained are listed in tables 7.9 and 7.10 for field trials 92 and 95, respectively.

### 7.3.3 *Dry matter weight*

Reflectance measurements do not correlate directly with dry matter weight, although they do correlate directly with soil cover and LAI. Nevertheless, the results of dry matter weight for field trials 116 in 1982 and 1983 are presented in appendix 15. The CV values are again large, but not as large as for LAI, especially at the end of the season. Because of the large CV values, the significance of treatment effects was not stable throughout the season.

### 7.3.4 *Other observations during the growing season*

In field trial 116 in 1982 some lodging occurred in the plots of the early-sown crop receiving the two highest levels of nitrogen. This caused some decrease in dry matter weight (appendix 15), but showed no distinct influence on LAI. In no other field trial lodging occurred.

From appendix 11 it can be concluded that during the 1982 season the soil was quite dry during missions up to mid-June (complete soil cover with barley), because there was little rainfall on days prior to the missions. At the beginning of the 1983 growing season, at low soil cover values with barley, considerable differences in soil moisture content occurred between missions. On 16 April and 6 May the soil surface was dry, on 30 April and 30 May the soil surface was wet because of rainfall on the preceding days.

In appendix 16 the time of occurrence of some development stages (recorded in the Feekes scale) are given for field trial 116 in 1982 and in 1983, respectively. The time of occurrence of some development stages for field trial 92 in 1982 and field trials 92 and 95 in 1983 are given in appendix 17.

## 7.4 Reflectance factors

For one field trial (trial 116 in 1982) results are presented in detail in the main text by means of figures and tables. Since LAI is the main plant characteristic to be estimated within this monograph and, as will be shown later in this section (see also chapter 4), the infrared reflectance is the reflectance that relates best to LAI, only figures and tables of the infrared reflectance will be presented in the main text of this section for field trial 116 in 1983, which was investigated in detail for estimating soil cover and LAI. The results of the green and red reflectances are given in appendices. The results of reflectance measurements for the field trials with wheat are also presented in appendices in order to limit the number of figures and tables within the main text.

In field trial 116 in 1982 reflectance factors were ascertained on 9 dates in the growing season by means of multispectral aerial photography in green, red and infrared passbands, as described in chapter 3. The effects of different treatments per date are illustrated in figure 7.6 for all three passbands.

The patterns of green and red reflectances were similar. In the first part of the season, reflectance in the visible passbands was lower for the early-sown crop because soil cover was higher than for the late-sown crop. At complete soil cover the colour of the vegetation was darker for the early-sown crop than for the late-sown crop. However, in the early-sown crop senescence also started

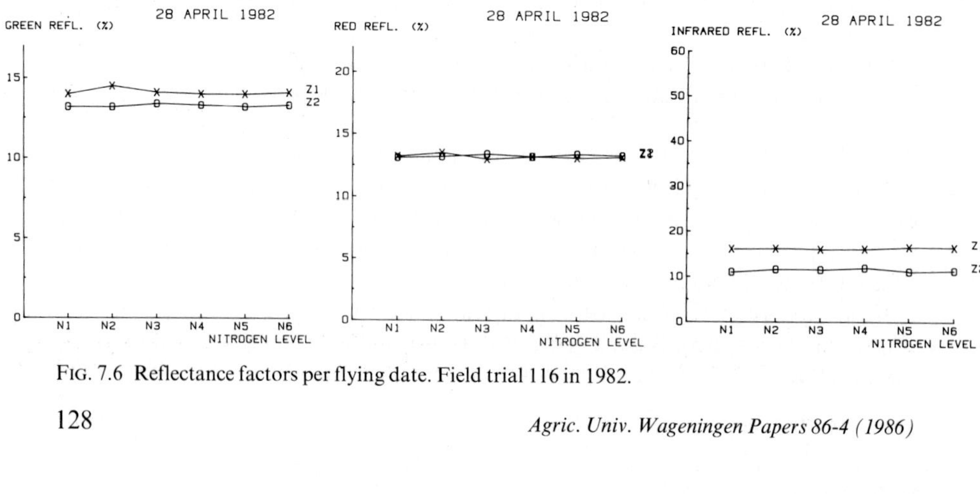


Fig. 7.6 Reflectance factors per flying date. Field trial 116 in 1982.

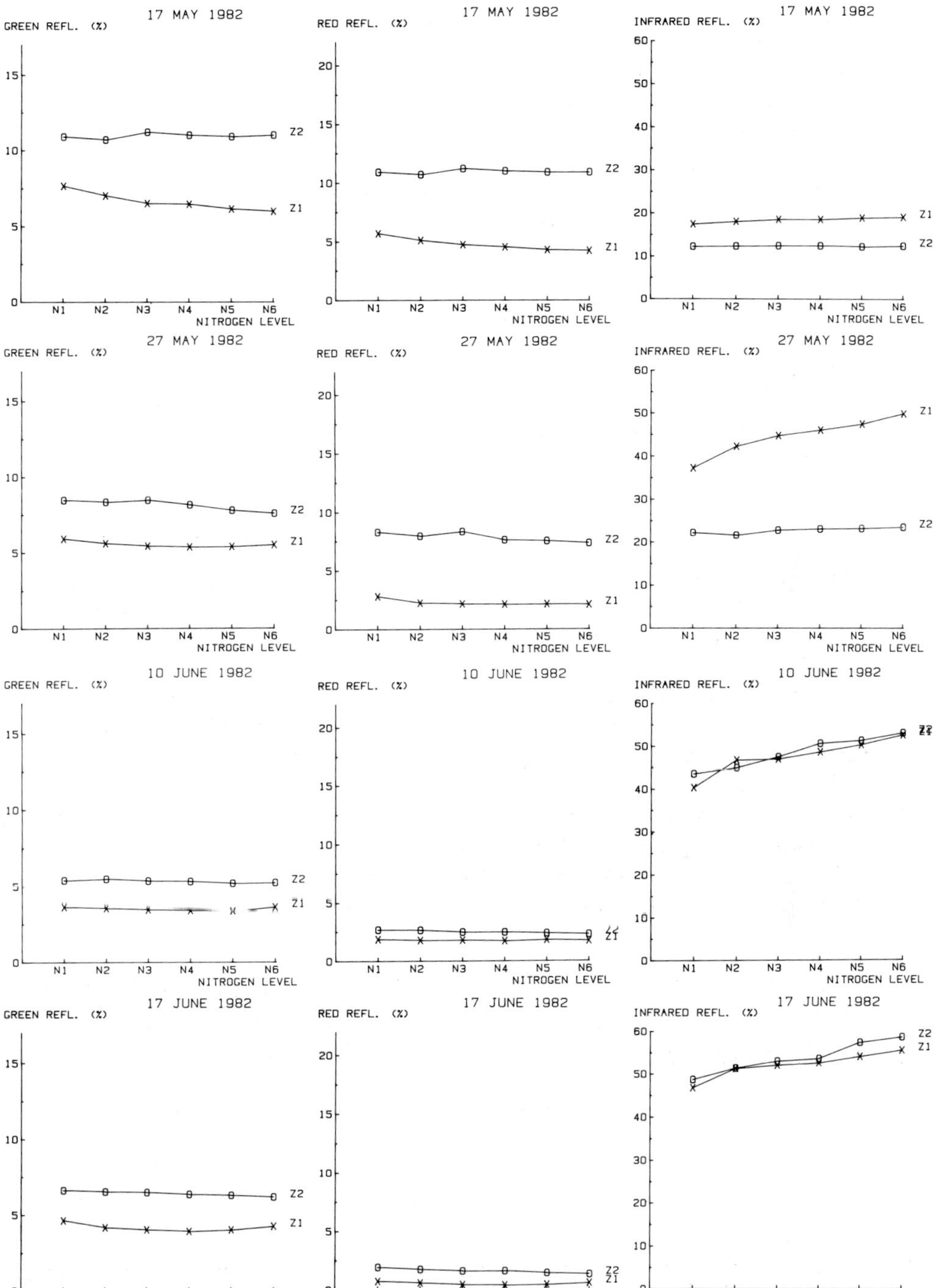


FIG. 7.6 continued.

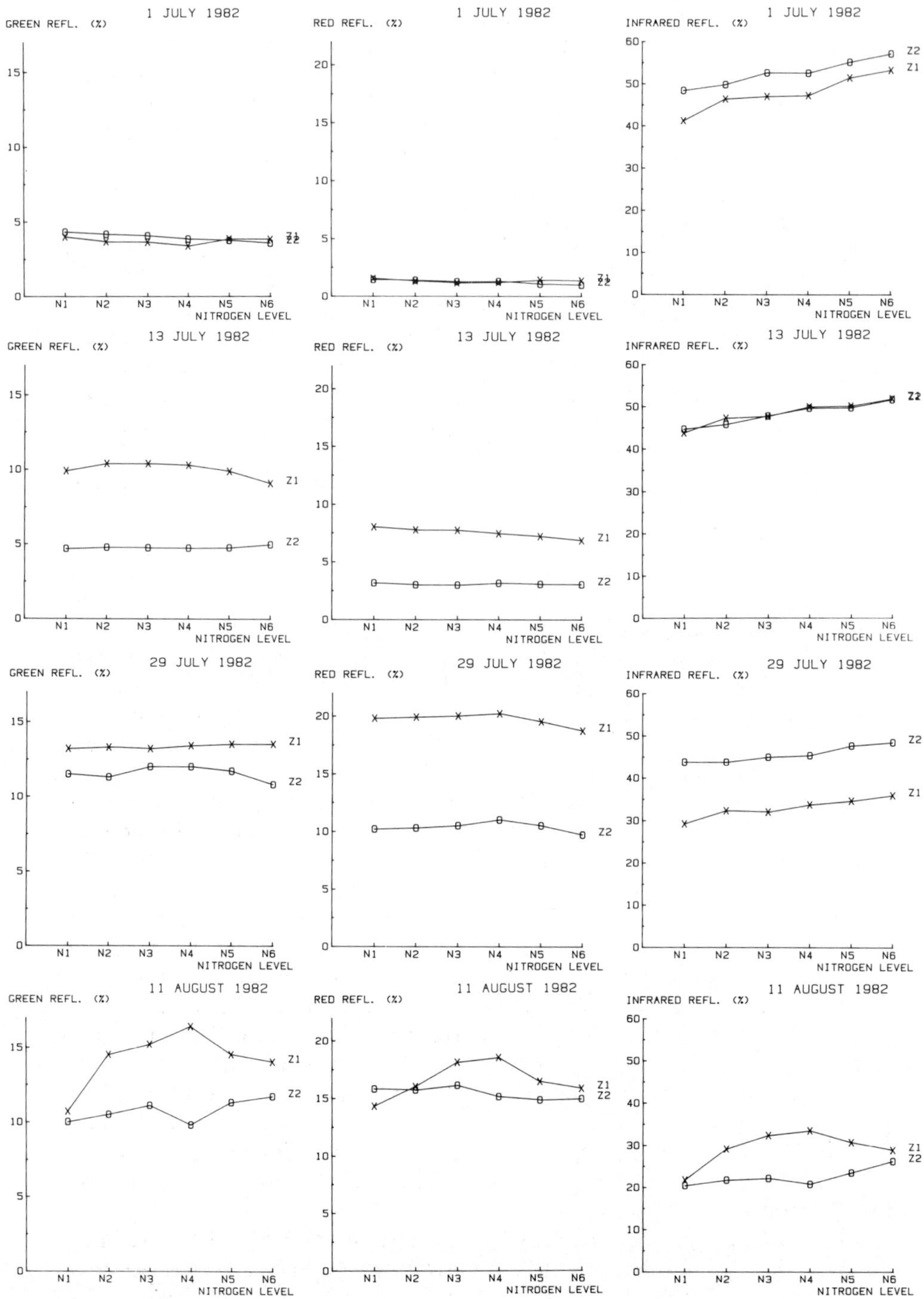


FIG. 7.6 continued.

earlier, resulting in higher reflectances in the visible passbands at the end of the season. Sowing date effects were evident nearly throughout the season, except at the beginning of July, where intersections occurred (tables 7.11 and 7.12). There were also significant (negative) nitrogen effects on reflectances in the visible passbands, but the differences were small. In the early-sown crop in particular there was a negative effect of nitrogen at the beginning of the season. Later, this effect became positive at the highest nitrogen levels, because of lodging.

At the beginning of the growing season infrared reflectance was, of course, higher for the early-sown crop. During the first half of June the late-sown crop caught up. These observations agree with the results obtained for LAI (figure 7.2). There was a positive effect of nitrogen on infrared reflectance in both crops throughout the season. This is also in agreement with the LAI results. It is also noteworthy that no influence of lodging was observed on the reflectance data in the infrared passband, except perhaps at the end of the season, when vegetation was dead. The same phenomenon was also noticed for the LAI.

Finally, in figure 7.7 the reflectances per sowing date in the three passbands are given as a function of time (days after sowing). The same features as noticed before are apparent from these figures.

In field trial 116 in 1983 reflectances were ascertained on 10 dates during the growing season. Treatment effects per date are illustrated in figure 7.8 for the infrared passband and in appendix 18 for the green and red passbands.

In the first part of the season, reflectances in the visible passbands were lower for the early-sown crop. In the second part of the season the sowing date effect reversed. The intersection occurred somewhere at the very beginning of July. On most dates there was a negative nitrogen effect in the visible passbands (mostly small in absolute value).

Infrared reflectances were highest for the early-sown crop during the beginning of the season. This effect reversed in the first half of June. The moment of this reversal agrees with the results of the LAI (figure 7.4). Throughout the season there was also a positive nitrogen effect on infrared reflectance in crops belonging to both sowing dates. This is also in agreement with the results for the LAI. During the first part of the growing season the nitrogen effect consisted of the reflectance of N2, N3, N4 and N5 being higher than that of N1. After the supplemental nutrition was applied at Feekes stage 7, there was a positive effect of nitrogen at all levels in the early-sown crop. In this crop the small difference between the two highest levels (N4 and N5) was notable. In the late-sown crop the nitrogen effect was less pronounced than in the early-sown crop. Until the beginning of July, N3 and N5 were exceeded by N2 and N4. Then N3 and N5 caught up with N2 because of the supplemental nutrition in June. However, N5 never caught up completely with N4. The change in these differences was very slow during the season.

The nitrogen effects described before are illustrated in figure 7.9 for the infrared passband and in appendix 19 for the green and red passbands, respectively, per sowing date as a function of time. In addition to these effects, the influence

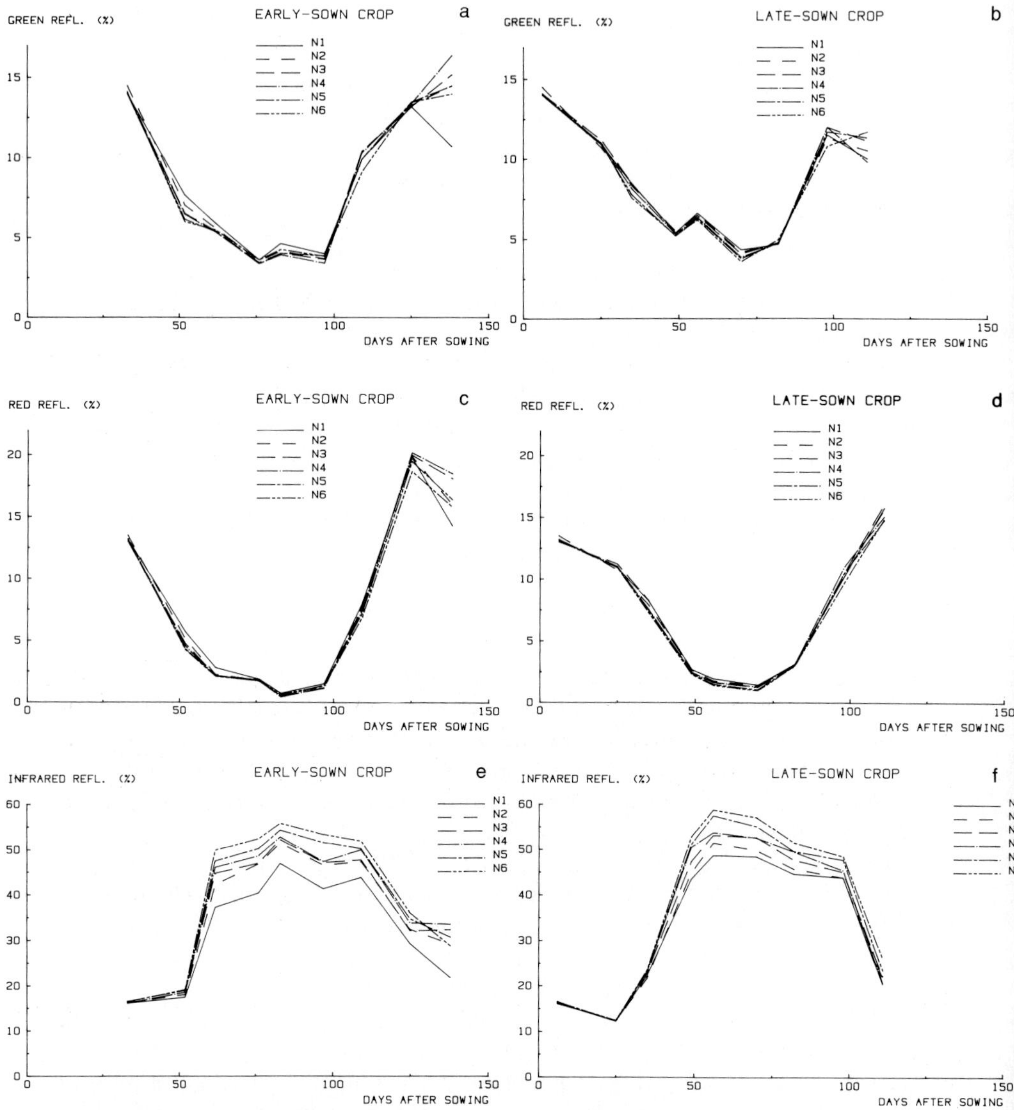


FIG. 7.7 Reflectance as a function of time. Field trial 116 in 1982.
Green reflectance for Z1 (a) and Z2 (b)
Red reflectance for Z1 (c) and Z2 (d)
Infrared reflectance for Z1 (e) and Z2 (f)

of soil moisture content at the beginning of the season is important. The influence of soil moisture content on reflectance was the same for all three passbands. Reflectances were relatively small at high soil moisture content (30 April and 30 May) and large at low soil moisture content (16 April and 6 May).

Finally, the decrease in reflectance towards the end of the growing season

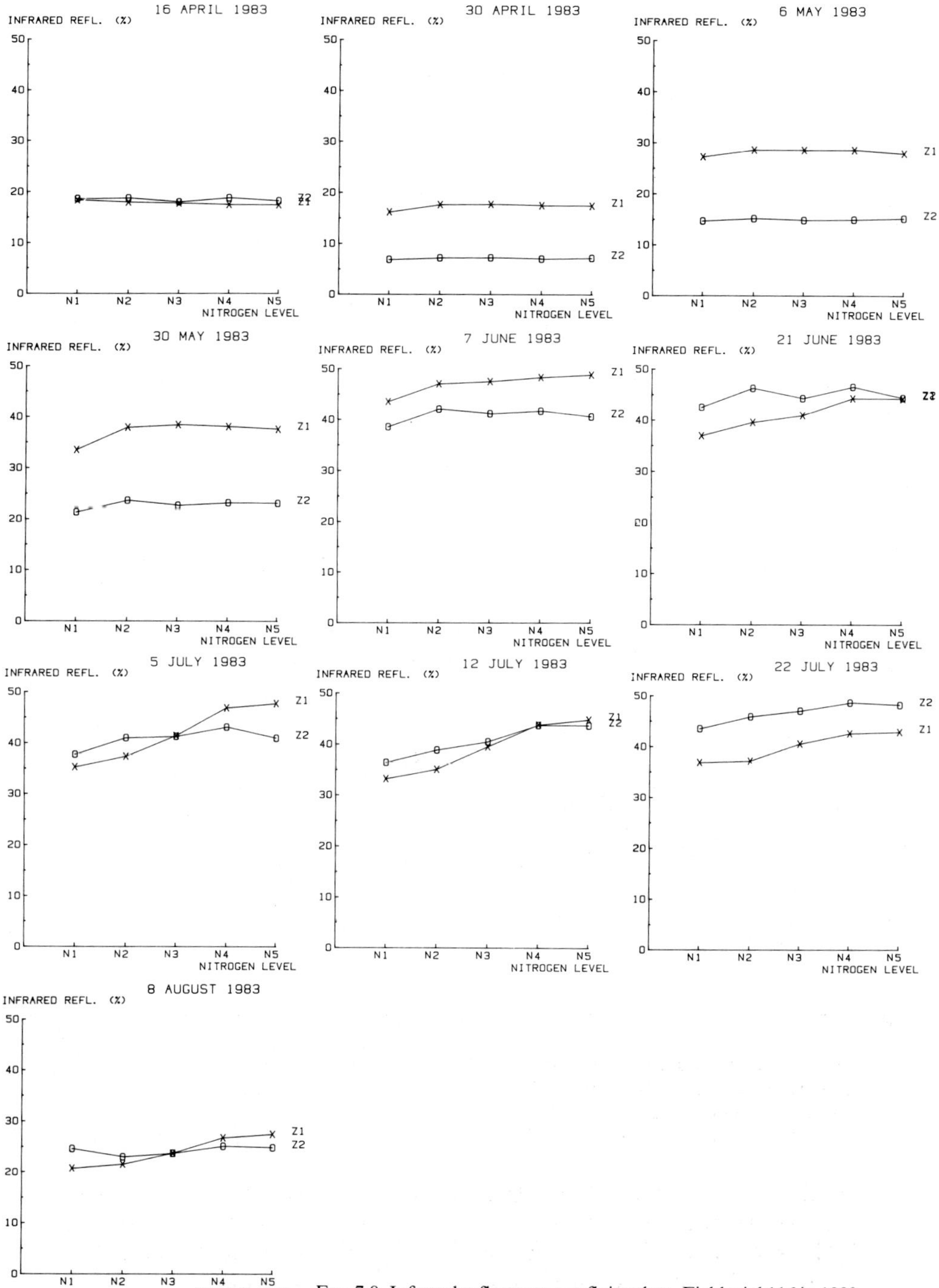


FIG. 7.8 Infrared reflectance per flying date. Field trial 116 in 1983.

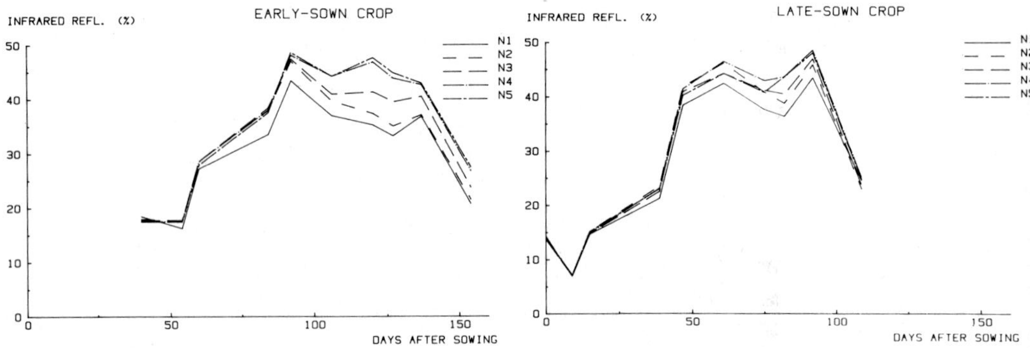


FIG. 7.9 Infrared reflectance as a function of time. Field trial 116 in 1983.

for the green and red passbands in the early-sown crop should be mentioned. This occurred for both the green and red reflectances with the same treatments, which almost rules out inaccuracies in measurement as the sole explanation. Moreover, the same phenomenon could be observed in field trial 116 in 1982 (figure 7.7). Probably an increased influence of soil background because of the wilting and even the dropping of dead leaves at the end of the growing season was responsible for this decrease in reflectance.

Treatment effects per date are illustrated in appendix 20 for 9 recording dates in field trial 92 in 1982. Results of reflectance measurements for 9 recording dates in field trials 92 and 95 in 1983 are illustrated in appendices 21 and 22, respectively.

For field trial 116 in 1982 the means, estimated coefficients of pooled variation and critical levels in testing for treatment effects are listed for the green, red and infrared passbands, respectively, in tables 7.11, 7.12 and 7.13.

TABLE 7.11 Green reflectance (%): means, CVs and critical levels in testing for treatment effects, obtained on consecutive missions. Field trial 116 in 1982 (36 plots).

| date | mean | CV | critical level in testing: | | |
|---|---|---|---|---|---|
| | | | inter-action | sowing dates | nitrogen nutrition |
| 28 April | 13.7 | 0.018 | 0.298 | 0.013 | 0.572 |
| 17 May | 8.78 | 0.046 | 0.000 | 0.007 | 0.000 |
| 27 May | 6.82 | 0.060 | 0.138 | 0.015 | 0.008 |
| 10 June | 4.41 | 0.040 | 0.548 | 0.001 | 0.215 |
| 17 June | 5.30 | 0.036 | 0.088 | 0.001 | 0.003 |
| 1 July | 3.84 | 0.075 | 0.130 | 0.124 | 0.059 |
| 13 July | 7.39 | 0.062 | 0.021 | 0.003 | 0.153 |
| 29 July | 12.4 | 0.036 | 0.107 | 0.012 | 0.319 |
| 11 August | 12.5 | 0.087 | 0.004 | 0.009 | 0.002 |

TABLE 7.12 Red reflectance (%): means, CVs and critical levels in testing for treatment effects, obtained on consecutive missions. Field trial 116 in 1982 (36 plots).

| date | mean | CV | critical level in testing: | | |
|---|---|---|---|---|---|
| | | | inter-action | sowing dates | nitrogen nutrition |
| 28 April | 13.2 | 0.022 | 0.286 | 0.602 | 0.884 |
| 17 May | 7.84 | 0.059 | 0.002 | 0.004 | 0.007 |
| 27 May | 5.03 | 0.088 | 0.321 | 0.003 | 0.007 |
| 10 June | 2.12 | 0.053 | 0.051 | 0.010 | 0.002 |
| 17 June | 1.10 | 0.113 | 0.038 | 0.001 | 0.001 |
| 1 July | 1.27 | 0.121 | 0.039 | 0.135 | 0.020 |
| 13 July | 5.27 | 0.072 | 0.018 | 0.005 | 0.010 |
| 29 July | 15.0 | 0.055 | 0.976 | 0.000 | 0.180 |
| 11 August | 16.0 | 0.185 | 0.233 | 0.722 | 0.244 |

TABLE 7.13 Infrared reflectance (%): means, CVs and critical levels in testing for treatment effects, obtained on consecutive missions. Field trial 116 in 1982 (36 plots).

| date | mean | CV | critical level in testing: | | |
|---|---|---|---|---|---|
| | | | inter-action | sowing dates | nitrogen nutrition |
| 28 April | 13.8 | 0.056 | 0.587 | 0.012 | 0.885 |
| 17 May | 15.4 | 0.021 | 0.003 | 0.001 | 0.001 |
| 27 May | 33.6 | 0.034 | 0.000 | 0.001 | 0.000 |
| 10 June | 47.9 | 0.071 | 0.864 | 0.465 | 0.001 |
| 17 June | 53.0 | 0.033 | 0.565 | 0.013 | 0.000 |
| 1 July | 50.2 | 0.033 | 0.361 | 0.011 | 0.000 |
| 13 July | 48.2 | 0.050 | 0.832 | 0.926 | 0.000 |
| 29 July | 39.3 | 0.042 | 0.507 | 0.006 | 0.001 |
| 11 August | 25.8 | 0.098 | 0.010 | 0.004 | 0.003 |

On most dates, critical levels in testing for treatment effects by means of reflectance measurements were smaller than those obtained by means of LAI measurements (table 7.6). This was especially true for the infrared reflectance, with the exception of large critical levels for sowing date effect on 10 June and 13 July, which resulted from a reversal. Thus treatment effects could be ascertained with larger power with reflectance measurements than with LAI determination in the field. This is the main reason for preferring reflectance measurements over sampling in field trials. This conclusion is also confirmed by the magnitude of the coefficients of variation. CV values in all three passbands were smaller than CV values for dry matter weight (appendix 15) and considerably smaller than CV values for LAI (table 7.6). Thus, spectral reflectance measurements obtained by means of multispectral aerial photography showed much smaller variability than field measurements (LAI and dry matter weight). This smaller variability probably occurred because large samples were used.

CV values were relatively large for the red passband in the middle of the growing season. During that period, red reflectances of the treatments were very low, resulting in measurements with relatively large variance (also, the atmosphere had a relatively large influence). The CV values of reflectance measurements of the mature crop (11 August) were relatively large too.

For the infrared passband in field trial 116 in 1983 the means, estimated CVs and critical levels in testing for treatment effects are listed in table 7.14. In appendix 23 the corresponding values for the green and red passbands are given. In this field trial, critical levels in testing for treatment effects by means of reflectance measurements, especially infrared reflectance (table 7.14), were again smaller than those obtained from field measurements of LAI (table 7.7). Also the variability (CV values) of reflectance measurements was much smaller than variability of field measurements (LAI and dry matter weight). The variability of the infrared reflectance was especially low. The significant sowing date effect on the earliest recording day (16 April) in the visible passbands when there was hardly any plant development is remarkable. It was primarily caused by a difference in soil tillage (the soil of the early-sown crop had been tilled, that of the late-sown crop had not). The significant nitrogen effect on the same day was, although significant, only small in absolute value and it primarily occurred in the early-sown crop (significant interaction effect).

The results of the analysis of variance for reflectance measurements of field trial 92 in 1982 and field trials 92 and 95 in 1983 are listed in appendices 24, 25 and 26, respectively. In these trials the same general conclusions can be drawn from comparing the results of reflectance measurements with those of LAI measurements as with field trials 116 in 1982 and 1983.

TABLE 7.14 Infrared reflectance (%): means, CVs and critical levels in testing for treatment effects, obtained on consecutive missions. Field trial 116 in 1983 (40 plots).

| date | mean | CV | critical level in testing: | | |
|---|---|---|---|---|---|
| | | | inter-action | sowing dates | nitrogen nutrition |
| 16 April | 18.1 | 0.050 | 0.464 | 0.356 | 0.209 |
| 30 April | 12.2 | 0.048 | 0.073 | 0.000 | 0.002 |
| 6 May | 21.6 | 0.041 | 0.024 | 0.000 | 0.002 |
| 30 May | 29.9 | 0.065 | 0.178 | 0.002 | 0.000 |
| 7 June | 43.8 | 0.044 | 0.453 | 0.001 | 0.004 |
| 21 June | 42.9 | 0.042 | 0.005 | 0.021 | 0.000 |
| 5 July | 41.2 | 0.049 | 0.000 | 0.193 | 0.000 |
| 12 July | 39.9 | 0.046 | 0.049 | 0.150 | 0.000 |
| 22 July | 43.4 | 0.029 | 0.048 | 0.002 | 0.000 |
| 8 August | 24.1 | 0.047 | 0.000 | 0.803 | 0.000 |

Soil cover in field trial 116 was ascertained on 4 dates in 1982 up to about the moment of complete cover (table 7.3). As noticed before there were no great differences in soil moisture content between those dates. Hence the reflectance value for bare soil in each passband could be considered as being approximately constant. Figure 7.1 shows the results of a measurement on bare soil with the field spectroradiometer (section 2.4) on the first recording date (28 April 1982). From these results the reflectance values of bare soil in the passbands used for multispectral photography can be estimated for this specific soil type. These estimates are: 12.4% in the green passband, 13.6% in the red passband, and 15.0% in the infrared passband (table 7.1).

The usefulness of different functions described in section 5.5 for estimating soil cover was checked and compared with the results for single passbands. These functions may be better estimators for soil cover than a single passband if soil moisture content varies strongly between dates with little soil cover. Because soil moisture content was similar on the early recording dates in 1982, any visible passband was expected to be superior in estimating soil cover than some function of differences or ratios of passbands.

As well as the reflectance of bare soil, the reflectance of vegetation in the different passbands has to be known if soil cover is to be estimated. The reflectance of vegetation may vary slightly because of differences in leaf colour, but the mean value (averaged over different dates and treatments) was used for each passband. Figure 7.7 and radiometer measurements (section 3.13) indicate that the reflectance of the barley crop in the green passband was approximately 4-5% and in the red passband approximately 2%. By using the inverse function of equation (5.16) for estimating the reflectance of bare soil and vegetation in the green and red passbands separately, the data from field trial 116 yielded the following estimates of the various reflectances (s = soil; v = vegetation; g = green; r = red):

$r_{s,g} = 12.4$, $r_{v,g} = 5.0$,

$r_{s,r} = 13.5$, $r_{v,r} = 2.0$.

These values agree very well with the values given earlier in this section. The values above were used as physical constants for checking the usefulness of different functions.

Table 7.15 presents a summary of the results of estimating soil cover with the single passbands and with the estimators discussed in section 5.5. For the ratio of red and green reflectance of bare soil on the experimental farm a value of 1.1 was found to be valid; this was indicated by the measurements with the field spectroradiometer (figure 7.1) as well as with aerial photography. The inverse value was used in equation (5.20).

As expected, the reflectance in a single passband was superior for estimating soil cover if no major differences in soil reflectance occurred. The red passband was preferable to the green passband.

TABLE 7.15 Results of different methods for estimating soil cover in field trial 116 in 1982.

| method | equation number | soil cover estimator | CV of residuals | figure |
|---|---|---|---|---|
| | 5.16 | $\frac{r_{s,g}-r_g}{r_{s,g}-r_{v,g}}$ | 0.109 | 7.10 |
| | 5.16 | $\frac{r_{s,r}-r_r}{r_{s,r}-r_{v,r}}$ | 0.069 | 7.11 |
| 1 | 5.20 | $\frac{r_g-C_1\cdot r_r}{r_{v,g}-C_1\cdot r_{v,r}}$ | 0.199 | 7.12 |
| 2 | 5.21 | $\frac{(r_{s,g}-r_{s,r})-(r_g-r_r)}{(r_{s,g}-r_{s,r})-(r_{v,g}-r_{v,r})}$ | 0.264 | 7.13 |
| 3 | 5.22 | $\frac{r_{s,g}-r_{s,r}\cdot r_g/r_r}{r_g/r_r\cdot(r_{v,r}-r_{s,r})-(r_{v,g}-r_{s,g})}$ | 0.271 | 7.14 |

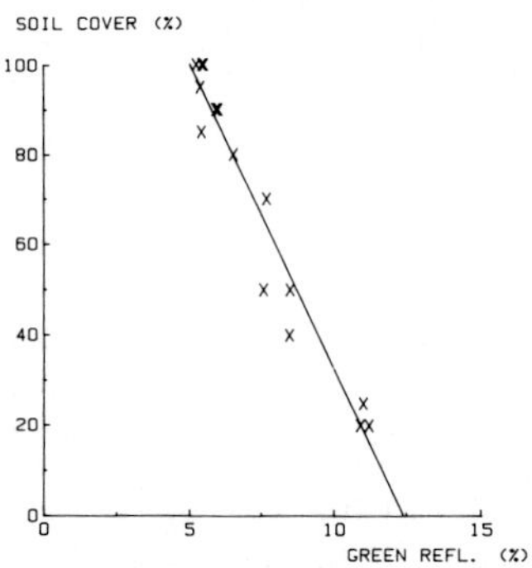


FIG. 7.10 Relationship between green reflectance and soil cover. Field trial 116 in 1982.

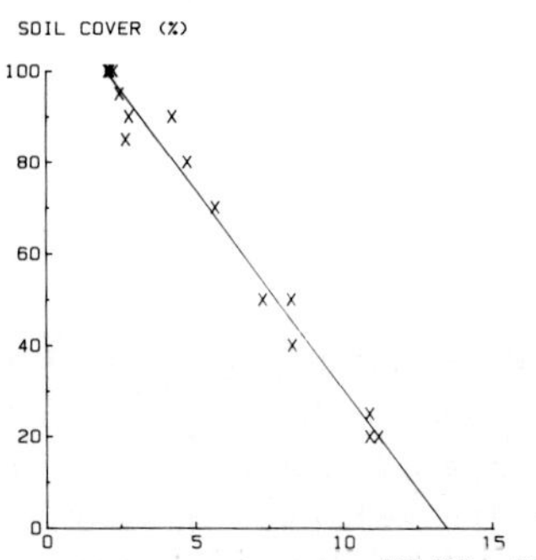


FIG. 7.11 Relationship between red reflectance and soil cover. Field trial 116 in 1982.

In field trials 116 and 100 in 1983, soil cover was ascertained on 6 dates until complete cover (tables 7.4 and 7.5) at all treatments. On 30 April no reflectances in the green passband were obtained; so results for this date had to be discarded.

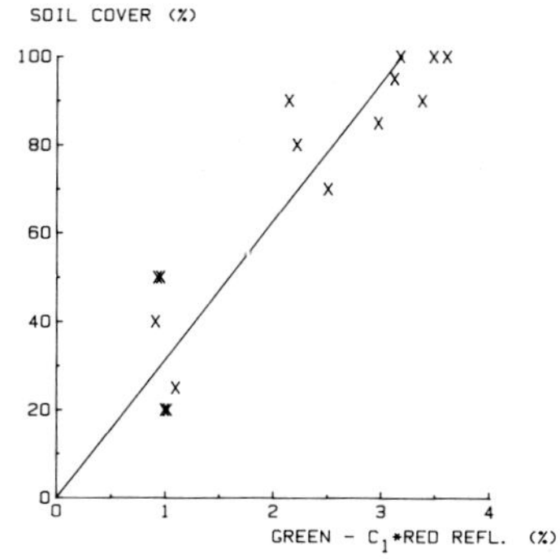


FIG. 7.12 Relationship between the function green-$C_1$.red and soil cover (method 1). Field trial 116 in 1982.

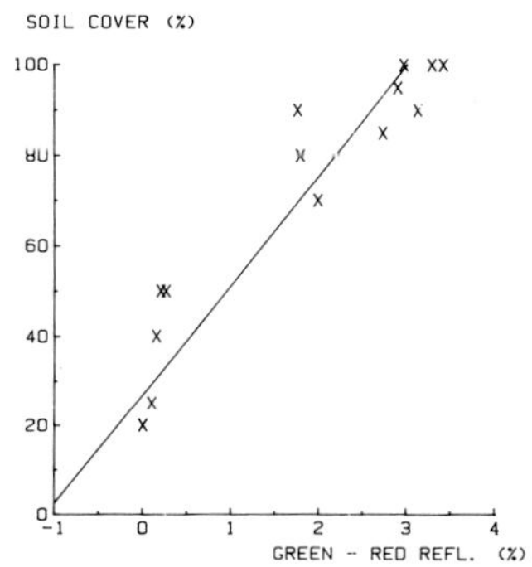


FIG. 7.13 Relationship between the green-red difference and soil cover (method 2). Field trial 116 in 1982.

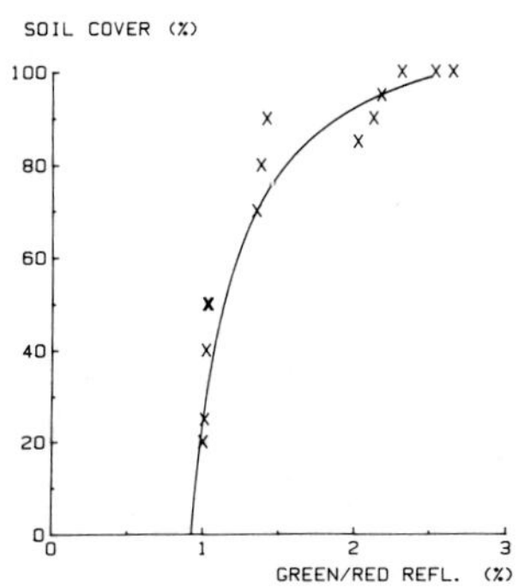


FIG. 7.14 Relationship between the green/red ratio and soil cover (method 3). Field trial 116 in 1982.

Results from field trial 100 were used for verifying relationships derived in section 5.5 or obtained in the first part of this section. Results of reflectance measurements of field trial 100 are listed in appendix 27.

Since large differences in soil moisture content at the beginning of the growing

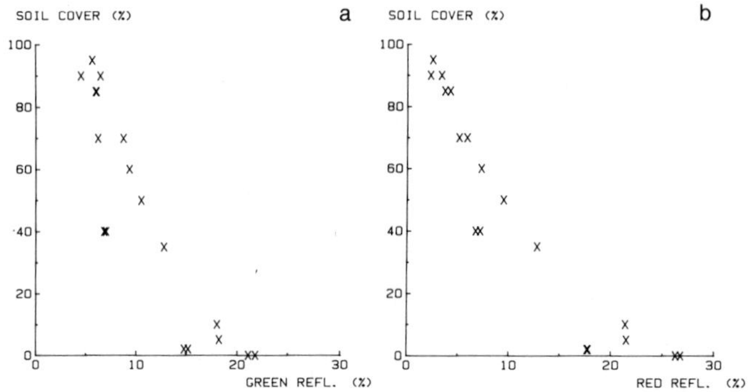


FIG. 7.15 Relationship between green (a) and red (b) reflectances and soil cover. Field trial 100 in 1983.

season caused large, but unknown differences in the reflectance of bare soil, a single passband was insufficient for estimating soil cover. This can be inferred from figure 7.15 for the green and red passbands in field trial 100. When the ratio of red to green reflectance of bare soil in this trial was calculated, it seemed to approximate a value of 1.2 instead of 1.1. For the reflectance of vegetation in the green and red passbands and for the reflectance of bare soil in the green passband the estimates obtained in field trial 116 in 1982 were used. For the reflectance of bare soil in the red passband a value of 14.9 (red/green ratio of 1.2) instead of 13.5 was used for estimating soil cover in field trial 116 in 1983 (resulting from the measurements in field trial 100). The results are presented in table 7.16 for the 3 methods described previously.

Theoretically, the best solution in correcting for differences in soil moisture content is provided by equation (5.20), since with equations (5.21) and (5.22) special assumptions were included for simplification (see section 5.5). However, as discussed in section 5.4, the inaccuracy may still be very large (CV = 0.525). Table 7.16 shows that the estimate based on the ratio of green and red reflectance

TABLE 7.16 Results of different methods for estimating soil cover in field trial 116 in 1983 ($r_{s,g} = 12.4$ and $r_{s,r} = 14.9$).

| method | equation number | soil cover estimator | CV of residuals | figure |
|---|---|---|---|---|
| 1 | 5.20 | $\frac{r_g - C_1 \cdot r_r}{r_{v,g} - C_1 \cdot r_{v,r}}$ | 0.525 | – |
| 2 | 5.21 | $\frac{(r_{s,g} - r_{s,r}) - (r_g - r_r)}{(r_{s,g} - r_{s,r}) - (r_{v,g} - r_{v,r})}$ | 0.436 | – |
| 3 | 5.22 | $\frac{r_{s,g} - r_{s,r} \cdot r_g/r_r}{r_g/r_r \cdot (r_{v,r} - r_{s,r}) - (r_{v,g} - r_{s,g})}$ | 0.317 | 7.16 |

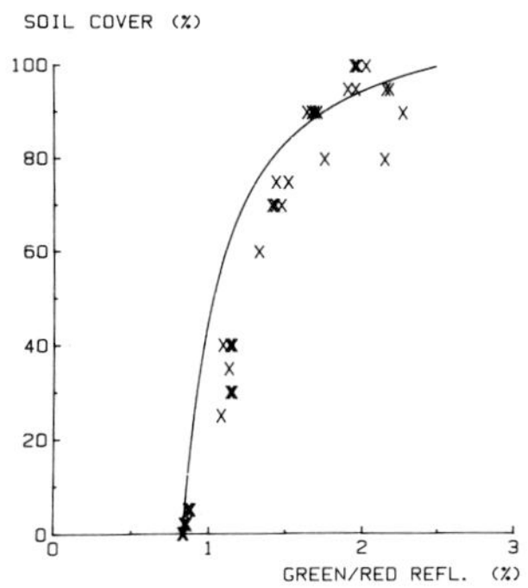


FIG. 7.16 Relationship between the green/red ratio and soil cover (for a red/green ratio of bare soil of 1.2). Field trial 116 in 1983.

gave the best results (CV = 0.317). This is illustrated in figure 7.16. This figure also shows how the steep slope of the curve could cause huge errors in estimating soil cover when soil cover is low. Thus, this ratio is only moderately suitable for estimating soil cover.

It should be noted that the poor results in estimating soil cover may partly be caused by inaccuracy in the method of ascertaining soil cover in the field. Since soil cover was not ascertained for replicates, no yardstick for this inaccuracy is available. However, results also indicate that inaccuracies in ascertaining the reflectances in visible passbands may considerably influence the result of any difference or ratio of visible passbands, as already discussed in section 5.4. Moreover, as explained in chapter 5, a definition of soil cover that differs from the one used in agronomy has to be applied in remote sensing studies. Therefore, it can be concluded that remote sensing techniques should primarily be used for estimating LAI rather than soil cover. In the next section it will be shown that the results of using remote sensing to estimate LAI, which is the more important crop characteristic in agronomy, were much better than those of estimating soil cover.

## 7.6 ESTIMATING LEAF AREA INDEX

Because of the large variability within the LAI data, it was decided to use the smoothed LAI data wherever possible for studying the relationship with reflectance measurements. The dates of the flights and of the field sampling did not coincide and therefore this smoothing also offered the possibility of interpolating the LAI values for the dates of flight missions.

No smoothing was carried out in field trial 100 in 1983, because a larger sample size was used and also the sampling dates coincided well with the dates on which reflectances were ascertained. In appendix 27 the results of LAI measurements in this field trial are listed together with the results of reflectance measurements. For sampling dates 12 and 22 July no LAI data were available for the early-sown crop.

In field trials 92 and 95 in 1983 too few LAI data were gathered in the field to enable smoothing to be done. Too few LAI data on the generative stage were available for ascertaining a regression curve of LAI on reflectance factors. Thus in these two trials only the vegetative stage was analysed, using data from missions flown near the harvesting dates.

*7.6.1 Vegetative stage*

Let us first consider the vegetative canopy only. The vegetative stage of cereals ends after the appearance of the last leaf; at this point the ear is about to appear and senescence will soon begin. This moment also coincides with maximum LAI. Appendix 16 and figure 7.2 indicate that for field trial 116 in 1982 this moment occurred on about 15 June for the early-sown crop and on about 25 June for the late-sown crop. For this reason, 17 June was designated as the last date on which the vegetative stage was recorded in this trial.

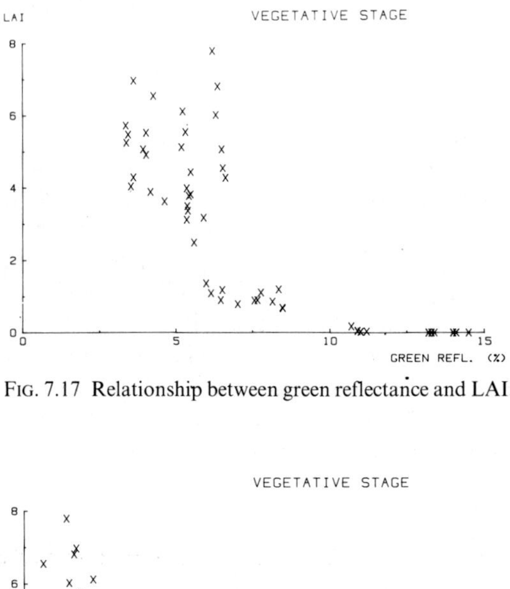


FIG. 7.17 Relationship between green reflectance and LAI. Field trial 116 in 1982.

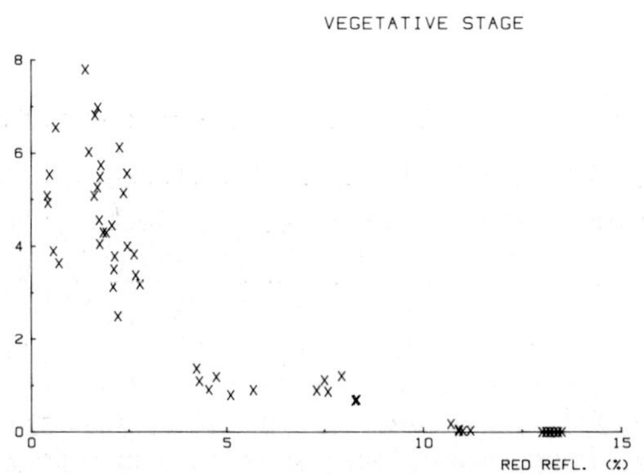


FIG. 7.18 Relationship between red reflectance and LAI. Field trial 116 in 1982.

The LAI is plotted against the green reflectance in figure 7.17. The reflectance decreased very rapidly with increasing LAI until LAI was about 2.0 (complete soil cover). At complete soil cover the reflectance was primarily determined by leaf colour and gave little information about LAI. The data were rather scattered. For the plot of LAI against the red reflectance a similar phenomenon was observed and this is illustrated in figure 7.18.

The LAI is plotted against the infrared reflectance in figure 7.19. Here, the reflectance increased with increasing LAI and this continued to be so at relatively large LAI levels.

After soil cover had been calculated by means of equation (5.16), using the red reflectance (section 7.5), the corrected infrared reflectance was calculated (equation 5.23 or 5.26). Finally, equation (5.12) was used for estimating LAI. The results of this estimation procedure are given in table 7.17.

Because soil moisture content was fairly constant between the early recording dates, the estimates for the reflectance of bare soil and vegetation given in section 7.5 were used. This also implies that the ratio $r_{s,ir}/r_{s,r}$ ($C_2$ in equation 5.25) was 1.1.

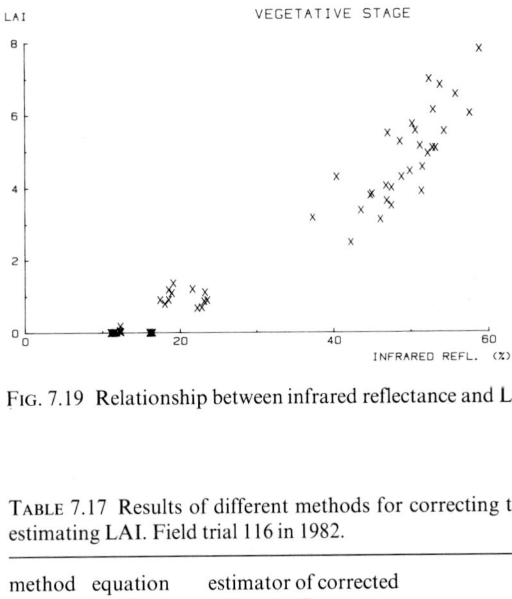


FIG. 7.19 Relationship between infrared reflectance and LAI. Field trial 116 in 1982.

TABLE 7.17 Results of different methods for correcting the infrared reflectance and subsequently estimating LAI. Field trial 116 in 1982.

| method | equation number | estimator of corrected infrared reflectance | $\hat{\alpha}$ | $\hat{r}_{\infty,ir}$ | CV of residuals | figure |
|---|---|---|---|---|---|---|
| 0 | 5.26 | $r_{ir} - r_{s,ir} \cdot \frac{r_r - r_{v,r}}{r_{s,r} - r_{v,r}}$ | 0.255 | 71.21 | 0.217 | 7.20 |
| 1 | 5.25 | $r_{ir} - \frac{C_2 \cdot (r_g \cdot r_{v,r} - r_r \cdot r_{v,g})}{C_1 \cdot r_{v,r} - r_{v,g}}$ | 0.228 | 75.70 | 0.230 | 7.21 |
| 2 | 5.28 | $r_{ir} - r_r$ | 0.252 | 68.57 | 0.214 | 7.22 |

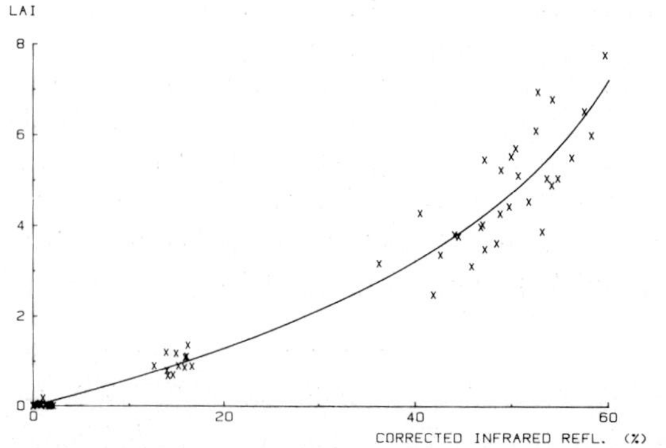


FIG. 7.20 Relationship between corrected infrared reflectance and LAI and the best fit according to method 0. Field trial 116, vegetative stage, 1982.

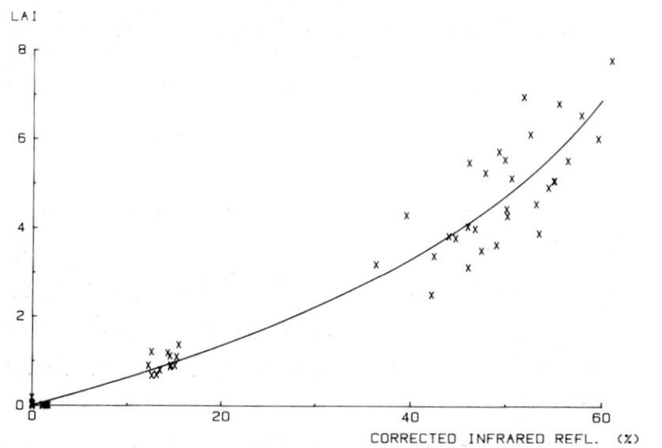


FIG. 7.21 Relationship between corrected infrared reflectance and LAI and the best fit according to method 1. Field trial 116, vegetative stage, 1982.

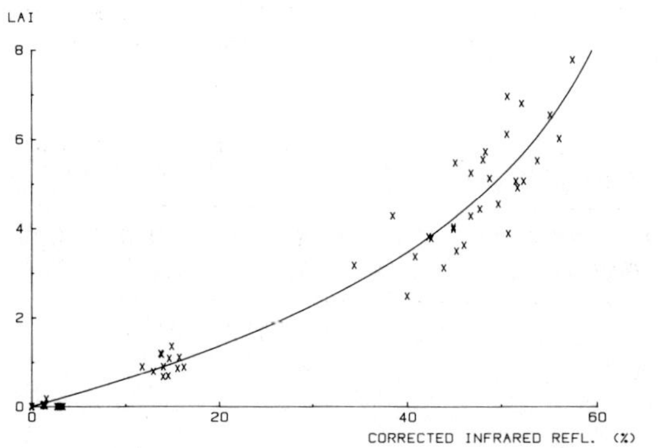


FIG. 7.22 Relationship between corrected infrared reflectance and LAI and the best fit according to method 2. Field trial 116, vegetative stage, 1982.

Since the ratio of infrared reflectance to red reflectance of the soil at the experimental farm did not differ greatly from the value 1.0, equation (5.28) may give a good approximation of the corrected infrared reflectance (method 2 of section 5.6). A more accurate method of correcting the infrared reflectance for background, by using the actual ratios of bare soil reflectance in the various passbands, is given by equation (5.25) (method 1 of section 5.6). These latter two methods (1 and 2) may be superior to the first one if soil reflectance cannot be regarded as remaining constant between the recording dates, because neither of these methods explicitly require the reflectance values of soil.

Table 7.17 shows that both methods 0 and 2 gave similar results. This indicates that method 2 may be very useful if soil reflectance is not similar on different dates. The conclusion that the simplifications induced by method 2 did not yield worse results than methods 0 or 1 was caused by the fact that the estimation of LAI is empirical, so that the estimates of parameters are not identical. For example, $\hat{r}_{\infty,ir}$ is the asymptotic value of the corrected infrared reflectance, and it is estimated from the measurements. For most data sets, $\hat{r}_{\infty,ir}$ is an extrapolation of the data, and this results in considerable variation in estimated $r_{\infty,ir}$ values, but with an optimal fit to the data. The results obtained from method 1 were slightly worse than those of methods 0 and 2 (larger CV value). This may be because the green passband correlated strongly with the red passband.

Since in 1983 soil reflectance was not constant between different days on which missions were flown, method 1 or 2 should be applied to correct for soil background. The results obtained in field trials 100 and 116 in 1983 are summarized in table 7.18. Results obtained by using method 1 were worse than those obtained by method 2, whether or not a value of 1.2 was used instead of 1.1 for either $1/C_1$ or $C_2$ or both. Hence, it was decided to concentrate on method 2.

In field trial 92 in 1982 the results obtained for the vegetative stage by using method 2 are illustrated in figure 7.25. The estimates for the two parameters were: $\hat{\alpha} = 0.286$ and $\hat{r}_{\infty,ir} = 58.75$. The CV was 0.271.

LAI data for field trials 92 and 95 in 1983 were not smoothed and harvesting dates were related to the nearest flying dates, in order to get some idea of the shape of the curve. Results for field trial 92 in 1983 are illustrated in figure

TABLE 7.18 Results of different methods for correcting the infrared reflectance and subsequently estimating LAI in field trials 100 and 116 in 1983.

| method | field trial | equation number | estimator of corrected infrared reflectance | $\hat{\alpha}$ | $\hat{r}_{\infty,ir}$ | CV of residuals | figure |
|---|---|---|---|---|---|---|---|
| 1 | 100 | 5.25 | $r_{ir} - \frac{C_2 \cdot (r_g \cdot r_{v,r} - r_r \cdot r_{v,g})}{C_1 \cdot r_{v,r} - r_{v,g}}$ | 0.321 | 69.06 | 0.213 | – |
| 1 | 116 | 5.25 | | 0.311 | 75.91 | 0.370 | – |
| 2 | 100 | 5.28 | $r_{ir} - r_r$ | 0.335 | 64.66 | 0.198 | 7.23 |
| 2 | 116 | 5.28 | | 0.328 | 70.66 | 0.344 | 7.24 |

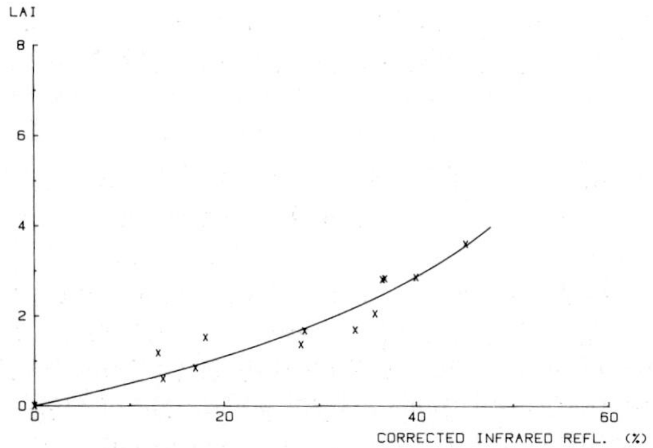


FIG. 7.23 Regression of LAI on corrected infrared reflectance (method 2). Field trial 100, vegetative stage, 1983.

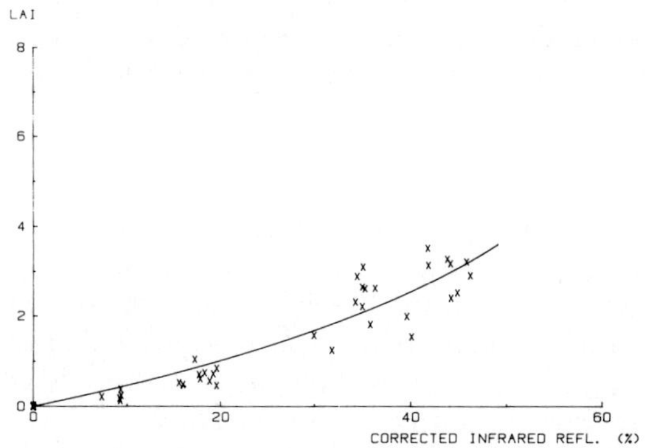


FIG. 7.24 Regression of LAI on corrected infrared reflectance (method 2). Field trial 116, vegetative stage, 1983.

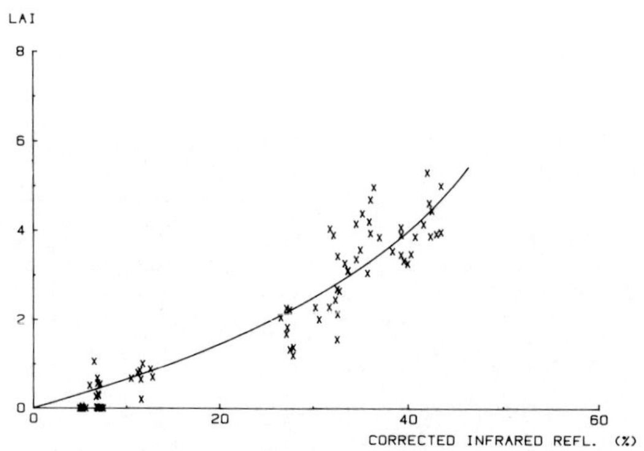


FIG. 7.25 Regression of LAI on corrected infrared reflectance (method 2). Field trial 92, vegetative stage, 1982.

7.26, with estimates $\hat{\alpha} = 0.393$, $\hat{r}_{\infty,ir} = 44.31$, and CV = 0.197. Results for field trial 95 in 1983 are illustrated in figure 7.27, with estimates $\hat{\alpha} = 0.402$, $\hat{r}_{\infty,ir} = 48.20$, and CV = 0.122.

The curves for both field trials 100 and 116 in 1983 (figures 7.23 and 7.24) correspond very well with one another. A conventional F test was applied to the two sets of data combined to test whether resulting curves differ significantly. The result of such a test for field trials 100 and 116 in 1983 at the vegetative stage was that the two curves were not significantly different. In that instance the data were from the same crop during the same growing season. The curve for field trial 100 in 1983 was just not significantly different at the 5% level from the curve obtained at field trial 116 in 1982. However, the curves for field trials 116 in 1982 and 1983 were significantly different. Results are plotted in figure 7.28. At this stage we have to conclude that regression curves for the same crop cannot be applied directly to different growing seasons, but that re-

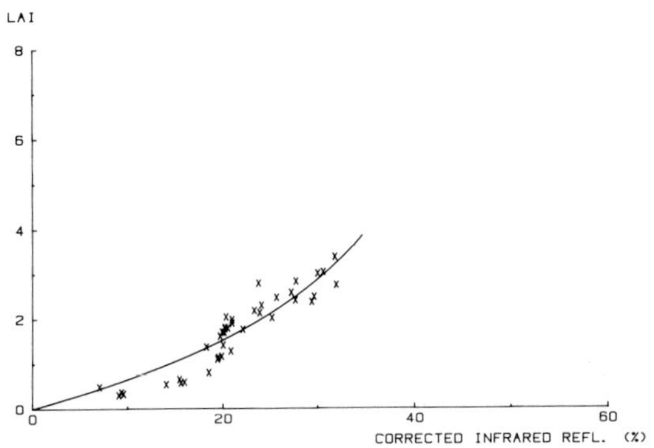


FIG. 7.26 Regression of LAI on corrected infrared reflectance (method 2). Field trial 92, vegetative stage, 1983.

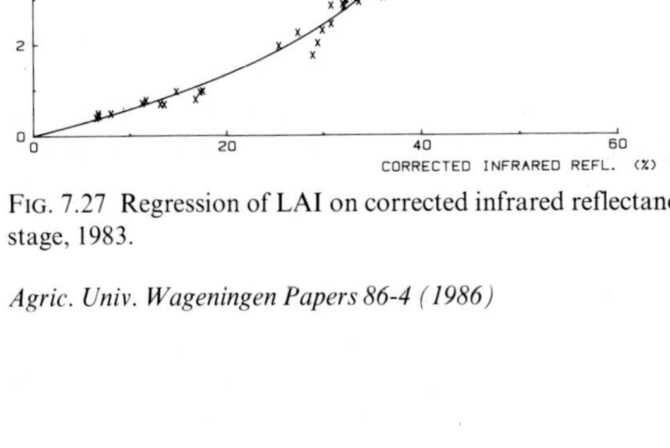


FIG. 7.27 Regression of LAI on corrected infrared reflectance (method 2). Field trial 95, vegetative stage, 1983.

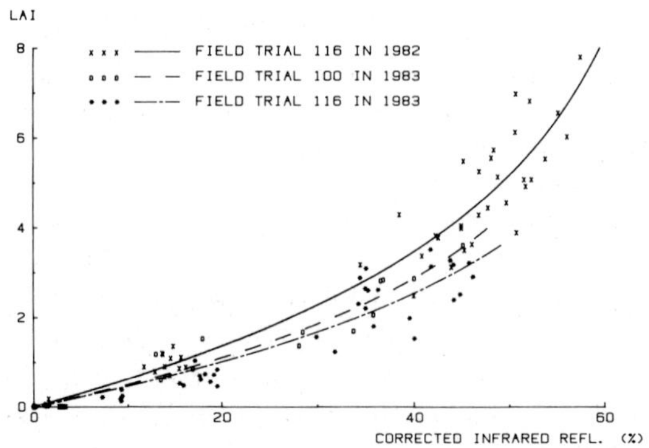


FIG. 7.28 Regression of LAI on corrected infrared reflectance (vegetative stage). Comparison of results from field trials 116 in 1982, 100 and 116 in 1983.

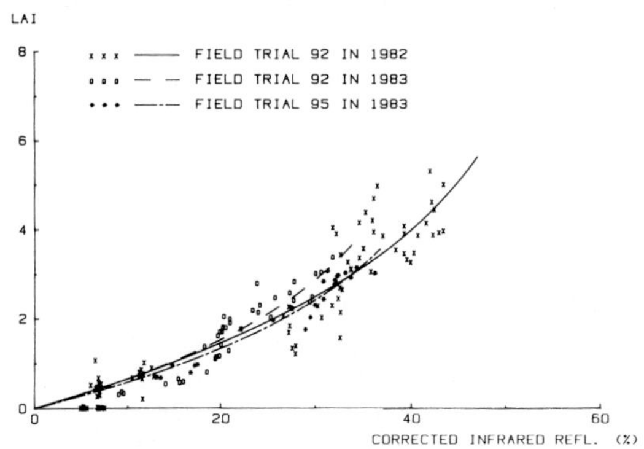


FIG. 7.29 Regression of LAI on corrected infrared reflectance (vegetative stage). Comparison of results from field trials 92 in 1982, 92 and 95 in 1983.

gression curves must be ascertained for different seasons separately, yielding different estimates of the parameters.

The curves for spring wheat in field trial 92 in 1982 and for winter wheat in field trials 92 and 95 in 1983 were significantly different from the curves for barley. The three curves for wheat at the vegetative stage were not significantly different at the 5% level (figure 7.29). However, it should be noted that the amount of data available on the trials with wheat in 1983 was small.

*7.6.2 Generative stage*

As stated in section 7.4, senescence at the end of the season caused an increase of the reflectance in the visible passbands and a decrease of the reflectance in the infrared passband in field trial 116 in 1983. At the end of the season, green reflectance returned to the level it had been at the beginning of the season with bare soil (about 15%) (figure 7.7). Red reflectance increased at the end of the season to a level that was higher than that of bare soil at the beginning of the season (about 20% at the end of the season). Infrared reflectance decreased at the end of the season to a value similar to the red reflectance, and was also higher than the infrared reflectance of bare soil. Results obtained by Ahlrichs & Bauer (1983) illustrate that the red and infrared reflectances of a wheat crop at maturity are very similar to those of bare soil.

In estimating LAI the measured infrared reflectance should be corrected for the background of yellow and dead leaves. For the situation with only bare soil and green vegetation, the difference between infrared and red reflectance (equation 5.28) appeared to be a good approximation for this corrected infrared reflectance (section 7.6.1). This presumably occurred because the reflectance of bare soil in the red passband did not differ greatly from that in the infrared passband. Because both reflectances also appeared to be nearly equal at the end of the season, the same equation may be used for correcting the infrared reflectance in this period. Then this equation would also be valid in the situation of some bare soil being visible within the canopy at the end of the season. Equa-

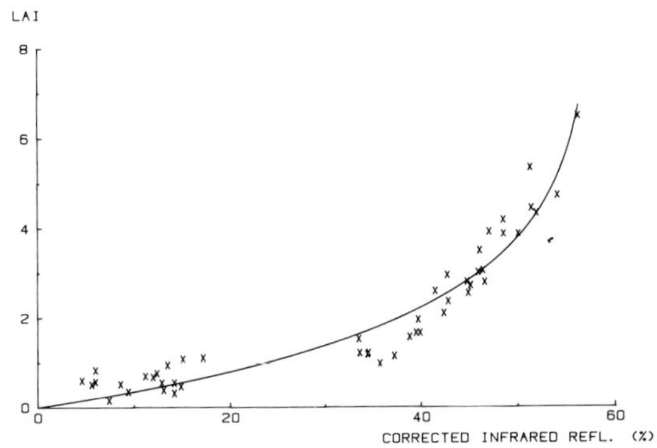


FIG. 7.30 Regression of LAI on corrected infrared reflectance (method 2). Field trial 116, generative stage, 1982.

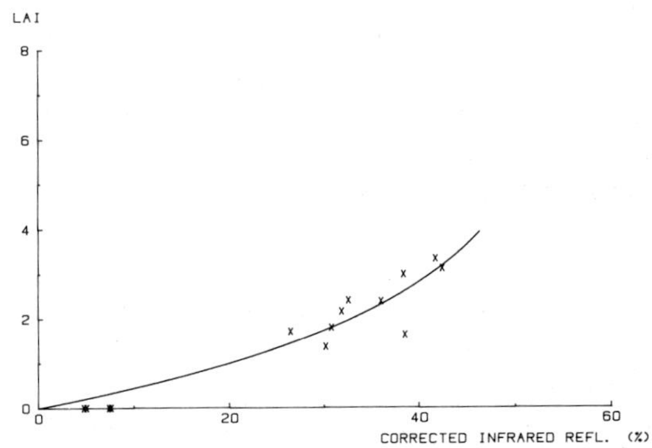


FIG. 7.31 Regression of LAI on corrected infrared reflectance (method 2). Field trial 100, generative stage, 1983.

tion (5.12) was again used for estimating the LAI from the corrected infrared reflectance. Because the crop structure at the generative stage will be different from that at the vegetative stage (at the generative stage ears will be present at the top of the canopy, while bare soil background will be replaced by yellow and dead leaves), the estimates of the two parameters will be different.

For the generative stage the regression of LAI on corrected infrared reflectance (by using equation 5.28) is illustrated in figure 7.30 for field trial 116 in 1982. The two parameters in equation (5.12) were estimated as: $\hat{\alpha} = 0.530$ and $\hat{r}_{\infty,ir} = 57.89$. The CV was 0.217, which is similar to the CV for the vegetative stage. As expected, the estimates of the parameters differed from those at the vegetative stage.

The regression of LAI on corrected infrared reflectance (using equation 5.28) is illustrated for field trial 100 in 1983 in figure 7.31. The estimates of the two parameters in equation (5.12) were: $\hat{\alpha} = 0.441$ and $\hat{r}_{\infty,ir} = 56.27$. The CV was 0.248.

The results for the generative stage of field trial 116 in 1983 are illustrated

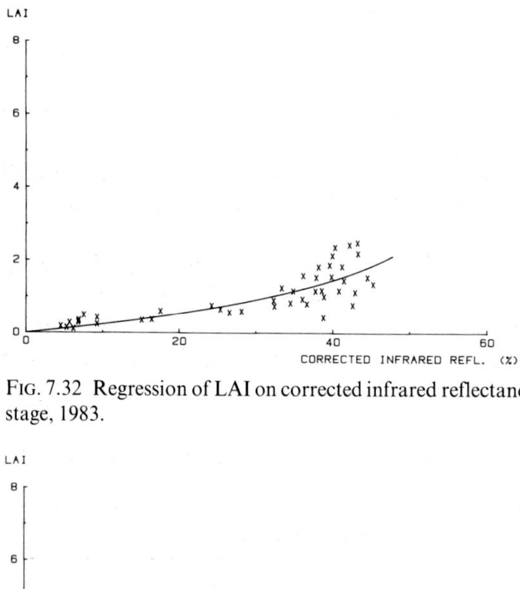


FIG. 7.32 Regression of LAI on corrected infrared reflectance (method 2). Field trial 116, generative stage, 1983.

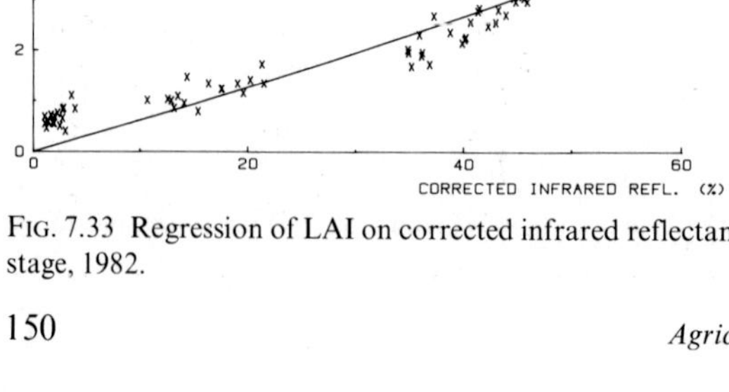


FIG. 7.33 Regression of LAI on corrected infrared reflectance (method 2). Field trial 92, generative stage, 1982.

in figure 7.32; the estimates were: $\hat{\alpha} = 0.763$ and $\hat{r}_{\infty,ir} = 59.49$. The CV was 0.394.

Results for the generative stage in field trial 92 in 1982 are illustrated in figure 7.33, with estimates $\hat{\alpha} = 0.074$, $\hat{r}_{\infty,ir} = 223.7$, and CV = 0.228. The extreme values for $\hat{\alpha}$ and $\hat{r}_{\infty,ir}$ illustrate the empirical nature of these parameters (the data in figure 7.33 did not show any asymptote, so the parameter describing the asymptotic value for the corrected infrared reflectance will have been estimated inaccurately. However, this does not influence the fit of the curve).

No regression curve was ascertained for field trials 92 and 95 in 1983 at the generative stage.

The regression curve for field trial 100 in 1983 resembles the one found for field trial 116 in 1982, although an F test shows that the two curves differ significantly. The curve for field trial 116 in 1983 is very different, because shortly after senescence started relatively low LAI values were measured. The latter may result from large subjectivity in classifying leaves as green or yellow (see section 1.2).

*7.6.3 Comparison of estimated LAI with measured LAI*

In section 7.4 the results of the analysis of variance for reflectance factors were compared with those for LAI (section 7.4.2). One has to keep in mind that the characteristics are different and therefore one should be careful when comparing these results.

With the results of section 7.6.1 and 7.6.2 it is possible to estimate the LAI per plot for the various field trials and subsequently to do an analysis of variance on these LAI estimates.

The means, CVs and critical levels of the estimated LAI in testing for treatment effects in field trial 116 in 1982 are listed in table 7.19. Results for 28 April 1982 are not listed, because bare soil predominated on that date. Similarly to the results for the infrared reflectance, the critical levels in testing for treatment

TABLE 7.19 LAI estimated by reflectance measurements: means, CVs and critical levels in testing for treatment effects, obtained on consecutive missions. Field trial 116 in 1982.

| date | mean | CV | critical level in testing: | | |
|---|---|---|---|---|---|
| | | | interaction | sowing dates | nitrogen nutrition |
| 17 May | 0.48 | 0.099 | 0.000 | 0.002 | 0.000 |
| 27 May | 2.42 | 0.075 | 0.000 | 0.001 | 0.000 |
| 10 June | 4.47 | 0.152 | 0.906 | 0.814 | 0.002 |
| 17 June | 5.70 | 0.073 | 0.279 | 0.110 | 0.000 |
| 1 July | 3.86 | 0.113 | 0.028 | 0.023 | 0.000 |
| 13 July | 2.63 | 0.101 | 0.809 | 0.138 | 0.000 |
| 29 July | 1.15 | 0.132 | 0.562 | 0.004 | 0.001 |
| 11 August | 0.36 | 0.402 | 0.122 | 0.189 | 0.022 |

effects in LAI estimates were again smaller than those in the LAI measurements (table 7.6) on most dates. Thus, treatment effects on LAI could be ascertained with larger power by reflectance measurements than by taking samples in the field. This was confirmed by the smaller CV values for the estimates of LAI. However, these latter CV values were larger than those for infrared reflectance.

Similar conclusions could be drawn for the other field trials. Results of the analysis of variance for field trial 116 in 1983 are listed in table 7.20. Those for field trial 92 in 1982 are listed in table 7.21. For field trials 92 and 95 in 1983 an analysis of variance was done on only some of the LAI estimates, corresponding to the LAI data available (tables 7.9 and 7.10). These results are listed in tables 7.22 and 7.23, respectively.

TABLE 7.20 LAI estimated by reflectance measurements: means, CVs and critical levels in testing for treatment effects, obtained on consecutive missions. Field trial 116 in 1983.

| date | mean | CV | critical level in testing: | | |
|---|---|---|---|---|---|
| | | | interaction | sowing dates | nitrogen nutrition |
| 6 May | 0.47 | 0.056 | 0.000 | 0.000 | 0.000 |
| 30 May | 1.43 | 0.120 | 0.110 | 0.002 | 0.001 |
| 7 June | 2.54 | 0.086 | 0.231 | 0.000 | 0.002 |
| 21 June | 2.13 | 0.090 | 0.030 | 0.001 | 0.000 |
| 5 July | 1.46 | 0.109 | 0.001 | 0.944 | 0.000 |
| 12 July | 1.21 | 0.089 | 0.284 | 0.010 | 0.000 |
| 22 July | 1.01 | 0.080 | 0.156 | 0.000 | 0.000 |
| 8 August | 0.16 | 0.113 | 0.051 | 0.022 | 0.000 |

TABLE 7.21 LAI estimated by reflectance measurements: means, CVs and critical levels in testing for treatment effects, obtained on consecutive missions. Field trial 92 in 1982.
(For explanation of A, B, C etc. see table 7.8)

| date | mean | CV | critical level in testing: | | | | | | |
|---|---|---|---|---|---|---|---|---|---|
| | | | ABC | AB | AC | BC | A | B | C |
| 28 April | 0.39 | 0.086 | 0.443 | 0.305 | 0.964 | 0.521 | 0.574 | 0.000 | 0.375 |
| 17 May | 0.60 | 0.090 | 0.586 | 0.022 | 0.191 | 0.912 | 0.142 | 0.000 | 0.145 |
| 27 May | 2.47 | 0.075 | 0.689 | 0.022 | 0.804 | 0.705 | 0.695 | 0.000 | 0.548 |
| 10 June | 3.13 | 0.063 | 0.078 | 0.112 | 0.324 | 0.319 | 0.030 | 0.000 | 0.281 |
| 17 June | 4.21 | 0.048 | 0.949 | 0.529 | 0.232 | 0.051 | 0.976 | 0.000 | 0.033 |
| 1 July | 3.09 | 0.041 | 0.800 | 0.136 | 0.211 | 0.120 | 0.315 | 0.000 | 0.940 |
| 13 July | 2.54 | 0.041 | 0.510 | 0.227 | 0.220 | 0.124 | 0.011 | 0.720 | 0.128 |
| 29 July | 1.02 | 0.057 | 0.038 | 0.612 | 0.159 | 0.319 | 0.000 | 0.000 | 0.000 |
| 11 August | 0.14 | 0.252 | 0.363 | 0.088 | 0.691 | 0.917 | 0.025 | 0.022 | 0.000 |

TABLE 7.22 LAI estimated by reflectance measurements: means, CVs and critical levels in testing for treatment effects, obtained on consecutive missions. Field trial 92 in 1983.
(For explanation of A, B, C etc. see table 7.8)

| date | treat-ments considered | mean | CV | critical level in testing: | | | | | |
|---|---|---|---|---|---|---|---|---|---|
| | | | | AB | AC | BC | A | B | C |
| 16 April | F1 | 1.02 | 0.103 | – | – | 0.208 | – | 0.000 | 0.258 |
| 30 April | F1 | 1.90 | 0.068 | – | – | 0.868 | – | 0.000 | 0.622 |
| 6 May | F1 | 1.65 | 0.075 | – | – | 0.819 | – | 0.460 | 0.720 |
| 7 June | S2 | 2.06 | 0.096 | – | 0.094 | – | 0.019 | – | 0.122 |
| 21 June | S2 | 2.90 | 0.196 | – | 0.348 | – | 0.179 | – | 0.958 |

TABLE 7.23 LAI estimated by reflectance measurements: means, CVs and critical levels in testing for treatment effects, obtained on consecutive missions. Field trial 95 in 1983.
(For explanation of A, B, C etc. see table 7.8)

| date | treat-ments considered | mean | CV | critical level in testing: | | | | | |
|---|---|---|---|---|---|---|---|---|---|
| | | | | AB | AC | BC | A | B | C |
| 16 April | F1 | 0.73 | 0.166 | – | – | 0.237 | – | 0.000 | 0.524 |
| 30 April | F1 | 1.65 | 0.097 | – | – | 0.503 | – | 0.000 | 0.747 |
| 6 May | F1 | 1.66 | 0.094 | – | – | 0.363 | – | 0.000 | 0.534 |
| 7 June | S2 | 2.93 | 0.167 | – | 0.375 | – | 0.293 | – | 0.774 |
| 21 June | S2 | 2.37 | 0.098 | – | 0.306 | – | 0.004 | – | 0.158 |

### *7.6.4 Using the results from field trial 100 for estimating LAI in field trial 116 in 1983*

The objective of field trial 100 in 1983 was to ascertain the regression of LAI on reflectance, and subsequently to apply this regression for estimating LAI in field trial 116. As noted in sections 7.6.1 and 7.6.2, this procedure may give satisfactory results at the vegetative stage, but will give large discrepancies at the generative stage. This is illustrated in figure 7.34. The LAI in field trials 100 and 116 was measured by different persons with slightly different instruments. Deciding which leaf is still green and which is not is very subjective, especially during senescence. This may result in such large systematic discrepancies that direct comparison between LAI values predicted by reflectance measurements and those measured in the conventional manner becomes highly suspect.

An analysis of variance was also carried out on the LAI values that had been estimated by using the regression curves per stage (vegetative and generative) obtained in field trial 100. Results of this analysis are listed in table 7.24. The conclusions concerning critical levels and CV values are similar to those obtained by the analysis of variance for LAI values that had been estimated by using the regression curves of field trial 116 in 1983 themselves (see table 7.20). Howev-

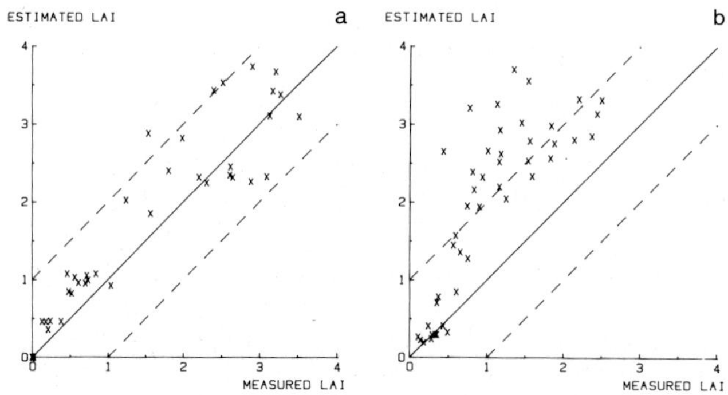


FIG. 7.34 LAI predicted by reflectance measurements plotted against the measured LAI at the vegetative (a) and generative (b) stage. A difference of 1.0 unit is indicated (---). Field trial 116 in 1983.

TABLE 7.24 LAI in field trial 116 in 1983 estimated by means of regression curves ascertained in field trial 100 in 1983: means, CVs and critical levels in testing for treatment effects.
A. calibration per stage (vegetative – generative)

| date of mission | mean | CV | critical level in testing: | | |
|---|---|---|---|---|---|
| | | | interaction | sowing dates | nitrogen nutrition |
| 6 May | 0.51 | 0.057 | 0.000 | 0.000 | 0.000 |
| 30 May | 1.58 | 0.126 | 0.099 | 0.002 | 0.001 |
| 7 June | 2.88 | 0.095 | 0.202 | 0.000 | 0.002 |
| 21 June | 3.01 | 0.096 | 0.004 | 0.090 | 0.000 |
| 5 July | 2.81 | 0.123 | 0.001 | 0.791 | 0.000 |
| 12 July | 2.31 | 0.094 | 0.373 | 0.010 | 0.000 |
| 22 July | 1.94 | 0.084 | 0.136 | 0.000 | 0.000 |
| 8 August | 0.30 | 0.113 | 0.050 | 0.022 | 0.000 |

B. calibration per mission date (and stage)

| date of mission | mean | CV | critical level in testing: | | |
|---|---|---|---|---|---|
| | | | interaction | sowing dates | nitrogen nutrition |
| 6 May | 0.83 | 0.057 | 0.000 | 0.000 | 0.000 |
| 30 May | 1.68 | 0.122 | 0.105 | 0.002 | 0.001 |
| 7 June | 2.80 | 0.090 | 0.219 | 0.000 | 0.002 |
| 21 June | 3.08 | 0.092 | 0.001 | 0.182 | 0.000 |
| 5 July | 2.95 | 0.115 | 0.001 | 0.879 | 0.000 |
| 12 July | 2.47 | 0.091 | 0.320 | 0.010 | 0.000 |
| 22 July | 1.30 | 0.081 | 0.150 | 0.000 | 0.000 |
| 8 August | 0.00 | – | – | – | – |

er, the new estimates of LAI were systematically different to the previous ones for the generative stage, possibly because different procedures were used to measure LAI (see above).

As well as ascertaining a regression curve per stage, regression curves were ascertained per recording date, in particular for field trial 100 in 1983. An analysis per recording date will result in fewer data points and a smaller section of the regression curve being covered than with an analysis per stage. The previously used multitemporal procedure (per stage) could be disturbed e.g. by (1) errors in atmospheric correction (see section 3.12), (2) variation in irradiation between film frames, (3) variation in leaf angle distribution between dates and (4) variation in sun elevation between dates. By using the curves of field trial 100, LAI was estimated in field trial 116 per plot and per date. Results are also listed in table 7.24. Since the results given in table 7.24B and in table 7.24A are similar, it may be concluded that the multitemporal analysis and the ascertainment of regression curves per stage was successful and that the disturbances enumerated before were negligible.

## 7.7 Field trial 85

As described in section 6.2.7, field measurements of a grass canopy were available for 18 May 1982. On 17 May 1982 aerial photographs were taken of this canopy in the green, red and infrared passbands. Since the dates were so close, the data may be compared with one another.

Treatment effects are illustrated in figure 7.35 for fresh weight, dry matter content, dry matter weight, nitrogen content and nitrogen weight, respectively. It was evident that nitrogen had a positive effect on fresh weight, dry matter weight and nitrogen weight, and on nitrogen content. The nitrogen effect on dry matter content was negative. Fresh and dry matter weights were best in rotation R3 and worst in rotation R2.

Treatment effects on reflectance measurements are illustrated in figure 7.36. The nitrogen effect on reflectances was as expected, namely negative in the green and red passbands and positive in the infrared passband. For the infrared reflectance, the rotation effect was similar to the effect on the weight variables. For the red and green reflectances it was opposite.

Results of an analysis of variance on all variables are listed in table 7.25. For nearly all variables there was a significant interaction effect (at the 5% level). This was most significant for the infrared reflectance and for the dry matter and nitrogen content. The small CV values for these variables are the reason why the significance of effects differs from that of the weight variables. The latter still have large CV values, in spite of the relatively large sample sizes (15 $m^2$).

The theory applied to cereals cannot be applied unmodified to grasses. With cereals the starting point is bare soil with zero crop weight. However, with

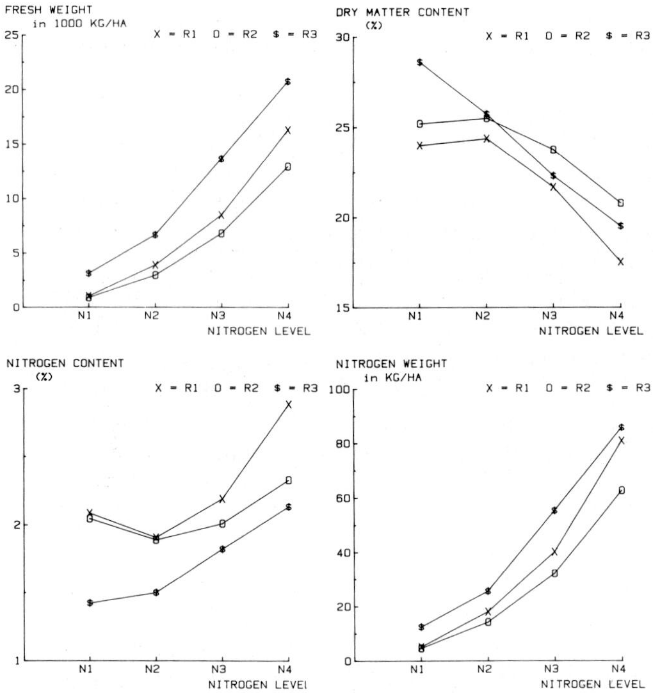


FIG. 7.35 Fresh weight (in kg/15 $m^2$), dry matter content (in %), dry matter weight (in 1000 kg/ha), nitrogen content (in %) and nitrogen weight (in kg/ha) for the grass canopy of field trial 85 on 18 May 1982.

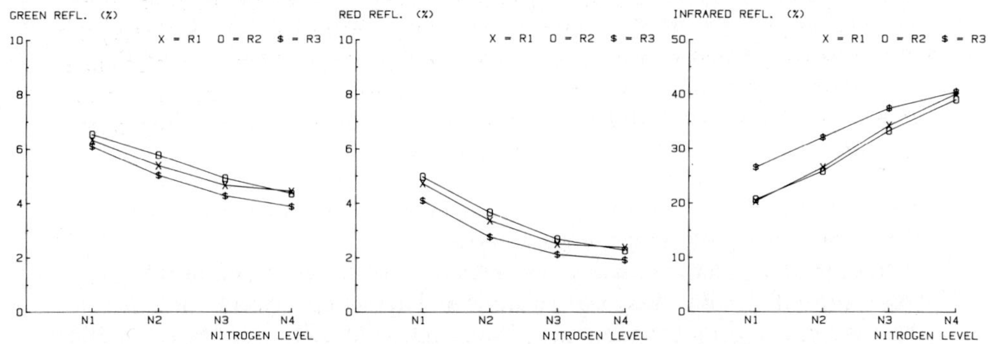


FIG. 7.36 Green, red and infrared reflectances for the grass canopy of field trial 85 on 17 May 1982.

TABLE 7.25 Reflectance in green, red and infrared passbands; fresh weight; dry matter content; dry matter weight; nitrogen content and nitrogen weight: means, CVs and critical levels in testing for treatment effects.
Field trial 85, 17/18 May 1982.

| | mean | CV | critical level in testing: | | |
|---|---|---|---|---|---|
| | | | interaction | rotation | nitrogen nutrition |
| green reflectance (%) | 5.13 | 0.038 | 0.154 | 0.002 | 0.000 |
| red reflectance (%) | 3.11 | 0.076 | 0.049 | 0.008 | 0.000 |
| infrared reflectance (%) | 31.4 | 0.055 | 0.000 | 0.023 | 0.000 |
| fresh weight (kg/ha) | 7500 | 0.186 | 0.016 | 0.000 | 0.000 |
| dry matter content (%) | 23.2 | 0.042 | 0.000 | 0.004 | 0.000 |
| dry matter weight (kg/ha) | 1740 | 0.147 | 0.040 | 0.000 | 0.000 |
| nitrogen content (%) | 2.1 | 0.059 | 0.000 | 0.001 | 0.000 |
| nitrogen weight (kg/ha) | 36.4 | 0.173 | 0.042 | 0.002 | 0.000 |

grasses the starting point is generally a stubble of mown or grazed grass, which is considered as being equivalent to zero crop weight. The crop weight is the weight of the harvested crop; some parts of the plant remain after harvesting. Since this stubble may partly consist of green plant material, the reflectance of this stubble in e.g. the red and infrared passbands will not be the same. This resulted in a shift of the abscissa of the regression curve at zero crop weight. It could not be tested whether the contribution of the stubble is constant in time, because data on this field trial were only available for one date and therefore no procedure could be developed to correct for background. Since all the measurements concerned a vegetative crop at about the same growth stage, it was possible to study a weight component instead of the LAI. An equation analogous to equation (5.12), but using the infrared reflectance instead of the corrected infrared reflectance and incorporating the zero shift, was used for estimating e.g. dry matter weight (DM) (in 1000 kg/ha):

$$DM = -1/\alpha \cdot \ln\left(1 - \frac{r_{ir} - \hat{r}_{st,ir}}{\hat{r}_{\infty,ir} - \hat{r}_{st,ir}}\right) \qquad (7.1)$$

with $\hat{r}_{\infty,ir}$: the asymptotic value for the infrared reflectance, and
$\hat{r}_{st,ir}$: the infrared reflectance of the stubble.

Fresh, dry matter and nitrogen weights yielded essentially similar results, although the best fit was for dry matter weight. The result for dry matter weight is illustrated in figure 7.37. The estimates of the parameters in equation (7.1) were: $\hat{\alpha} = 0.172$, $\hat{r}_{\infty,ir} = 72.77$ and $\hat{r}_{st,ir} = 18.37$. The CV was 0.236.

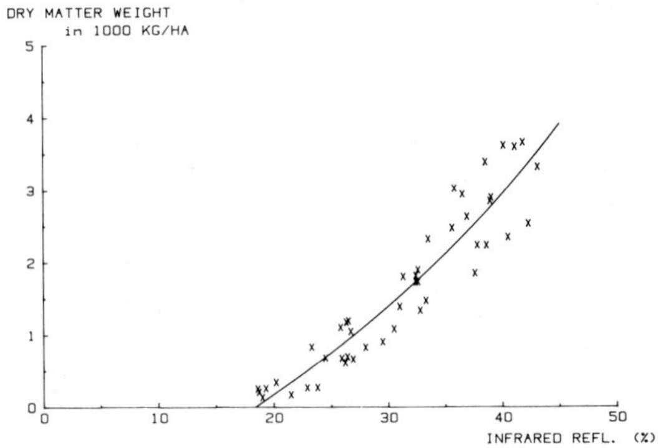


FIG. 7.37 Regression curve of dry matter weight on infrared reflectance. Field trial 85 on 17/18 May 1982.

## 7.8 FIELD TRIAL 87

As described in section 6.2.8, in this field trial with potatoes few field measurements were taken during the growing season. On 21 June 1982, 6 plants per plot were harvested for ascertaining the weight of foliage and tubers at that moment. These results were compared with aerial photography of the same field trial obtained on 17 June 1982. However, given the small number of plants harvested per plot, the accuracy of the field measurements was expected to be limited. The whole plots were harvested on 20 September and tuber yield was ascertained. This yield was the resultant of the crop growth during the whole growing season. An attempt was made to relate differences in tuber yield to differences in reflectance during the growing season, especially those in the infrared passband.

First of all, the results of field measurements taken on 21 June and aerial photography done on 17 June were compared. The potato crop predominantly consisted of green foliage at that time. Results for treatment effects are illustrated in figure 7.38. Results of an analysis of variance for field measurements

TABLE 7.26 Foliage and tuber weight on 21 June 1982, and tuber yield, fresh and dry, at the end of the growing season: means, CVs and critical levels in testing for treatment effects. Field trial 87 in 1982.

| | mean (g/m$^2$) | CV | critical level in testing: | | |
|---|---|---|---|---|---|
| | | | interaction | nematicide | rotation |
| foliage weight | 1577 | 0.141 | 0.218 | 0.315 | 0.354 |
| tuber weight | 387 | 0.253 | 0.901 | 0.044 | 0.378 |
| tuber yield (fresh) | 4132 | 0.053 | 0.149 | 0.684 | 0.001 |
| tuber yield (dry) | 1051 | 0.053 | 0.247 | 0.509 | 0.002 |

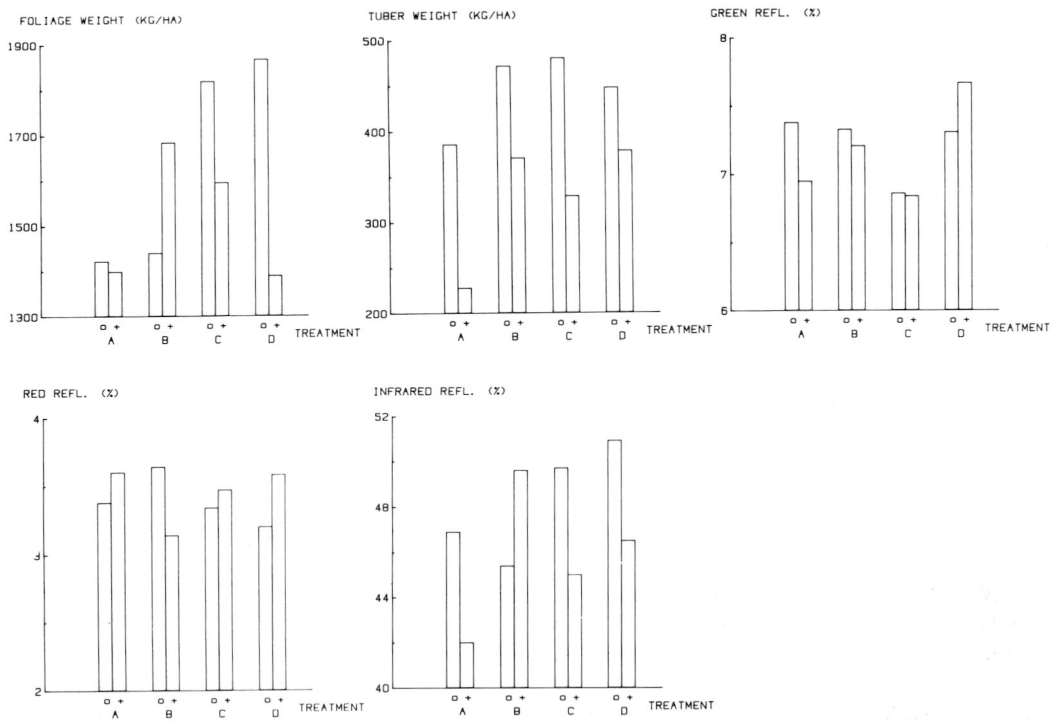


FIG. 7.38 Foliage and tuber weight (in kg/ha) on 21 June 1982, and green, red and infrared reflectances on 17 June 1982. Field trial 87 (for explanation of symbols see section 6.2.8).

(foliage and tuber weight) are listed in table 7.26 and for infrared reflectance are given in table 7.27. From the figures the relatively bad performance of the very narrow rotation (A) is evident. A negative effect of the nematicide treatment

TABLE 7.27 Reflectance (%) in the infrared passband on consecutive missions: means, CVs and critical levels in testing for treatment effects. Field trial 87 in 1982.

| date | mean | CV | critical level in testing: | | |
|---|---|---|---|---|---|
| | | | interaction | nematicide | rotation |
| 28 April | 11.1 | 0.056 | 0.384 | 0.061 | 0.601 |
| 17 May | 12.6 | 0.080 | 0.424 | 0.867 | 0.792 |
| 27 May | 15.5 | 0.066 | 0.997 | 0.963 | 0.457 |
| 10 June | 41.0 | 0.045 | 0.035 | 0.024 | 0.086 |
| 17 June | 47.0 | 0.028 | 0.004 | 0.007 | 0.013 |
| 1 July | 70.1 | 0.023 | 0.008 | 0.904 | 0.017 |
| 13 July | 59.9 | 0.027 | 0.493 | 0.937 | 0.037 |
| 29 July | 48.5 | 0.065 | 0.542 | 0.124 | 0.035 |
| 11 August | 34.9 | 0.069 | 0.091 | 0.323 | 0.000 |

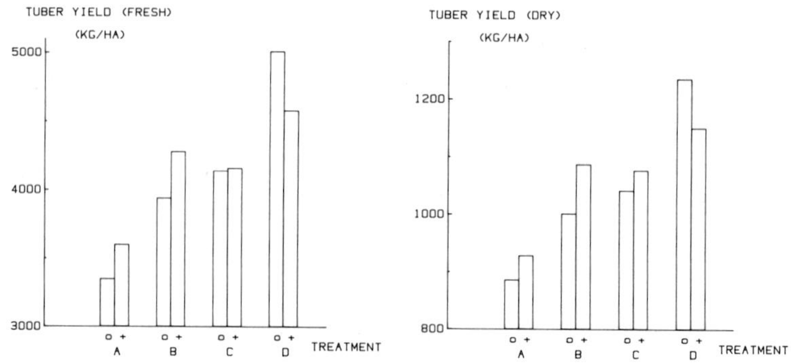


FIG. 7.39 Tuber yield (fresh and dry) in kg/ha in field trial 87 in 1982.

on tuber weight on 21 June is also apparent. Nematicide treatment also had a negative effect on foliage weight, except in rotation B, which exhibited a positive effect. This surprising result must be attributed to a phytotoxic action of the nematicide used ('Mocap'), at least at the beginning of the growing season. The effects on foliage weight are substantiated by the reflectance measurements, particularly by the infrared reflectance, which reveals a significant effect. Results for the green reflectance are less concurrent with the other results (figure 7.38).

A more accurate field measurement was the total yield of tubers at the end of the growing season. Results of treatment effects are illustrated in figure 7.39. Results of an analysis of variance on tuber yield are also listed in table 7.26. The relatively poor performance of rotation A is still evident. Contrary to the samples of 21 June, rotation D yielded the best results, whereas rotations B and C did not differ greatly. A slightly positive effect of the nematicide treatment on tuber yield appeared for rotations A, B and C. The effect for rotation D was still negative. Apparently, in the narrow rotations (A, B and C) the phytotoxic action of the nematicide used was overruled by the nematode-killing activity. Only for the 'healthy' rotation (D) was the nematode-killing activity minor, resulting in an overall negative effect.

The green, red and infrared reflectances are illustrated in figure 7.40 as a function of time. Since the infrared reflectance was expected to offer the most significant information about crop growth, only this reflectance was studied. A more explicit comparison of the two nematicide treatments per rotation is illustrated in figure 7.41 for the infrared reflectance. For rotation A, up to mid-July the infrared reflectance was lower in the plots with nematicide treatment than in those without nematicide treatment: thereafter this effect was reversed. For rotations C and D the nematicide effect seemed to be negative up to about mid-July and thereafter there was no clear nematicide effect. For rotation B the nematicide effect was positive throughout the season. The growing season could be roughly subdivided into two periods. Up to about mid-July most rotations exhi-

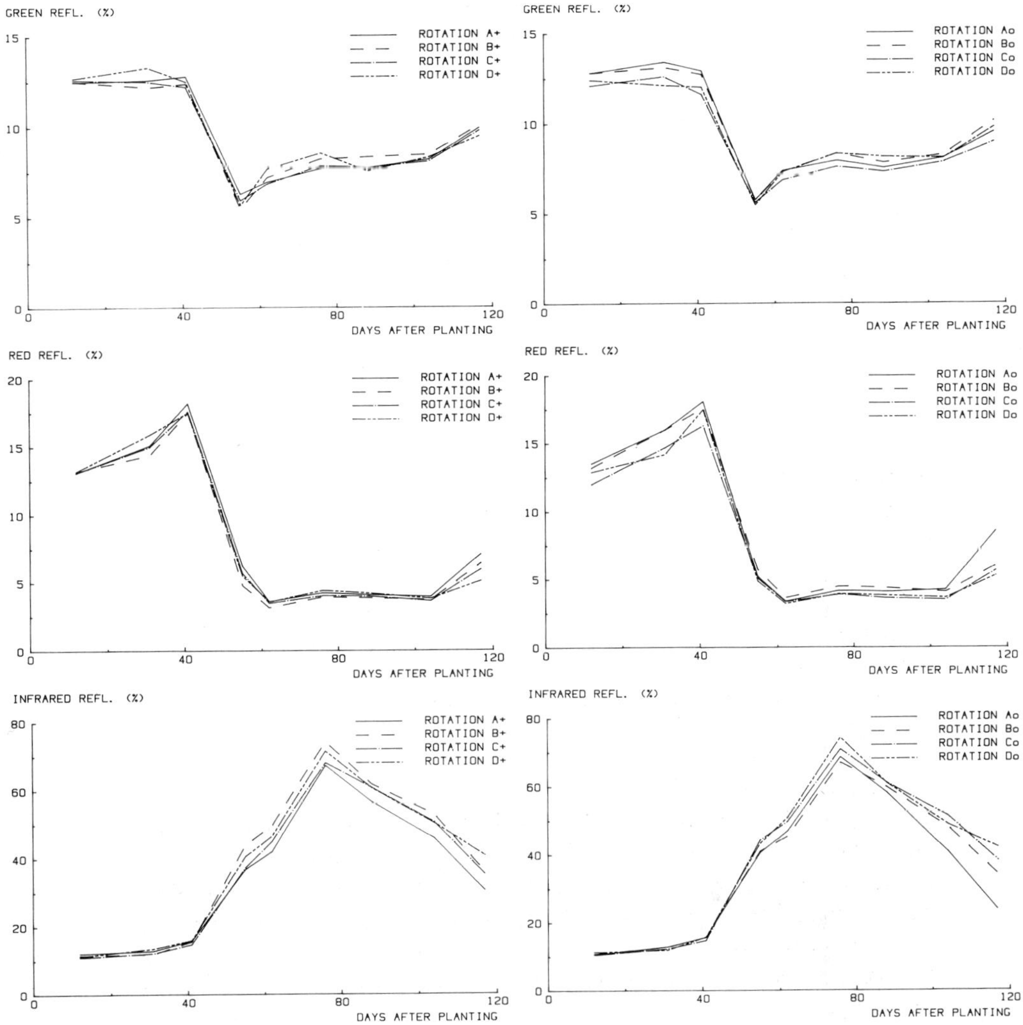


FIG. 7.40 Green, red and infrared reflectances as a function of time. Field trial 87 in 1982.

bited a negative nematicide effect. At the end of the growing season this negative effect was overruled or at least compensated for by a positive effect of the nematicide treatment. The latter period determined the speed at which the potato crop died off.

From figure 7.40 it is evident that the infrared reflectance was lowest for rotation A on nearly all dates. Thus, when differences in infrared reflectance were translated into differences in foliage weight or LAI and indirectly into differences in potential yield or tuber yield as such, the results indicated that the smal-

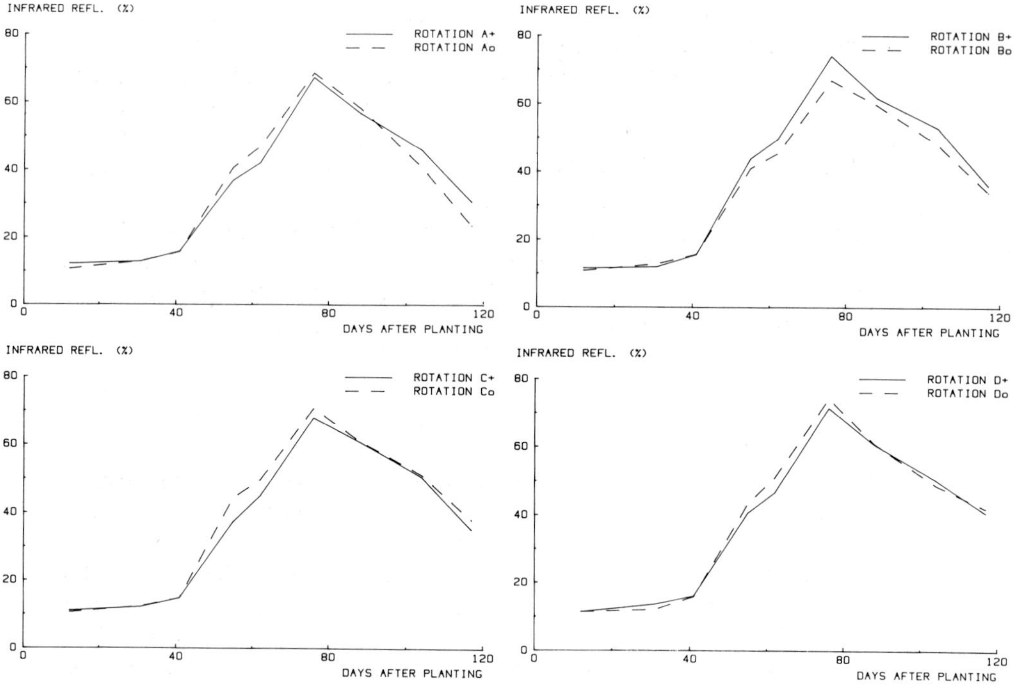


FIG. 7.41 Infrared reflectance as a function of time per rotation. Field trial 87 in 1982.

lest tuber yield would be found in rotation A. Actual tuber yield confirmed this (figure 7.39). The overall effect of the nematicide on tuber yield was positive, which means that the nematode-killing activity, expressed as a prolongation of the green crop, predominated. This agreed with the higher infrared reflectance at the end of the growing season for rotation A with nematicide treatment.

The positive effect of the nematicide treatment for rotation B during the whole growing season was also seen in the tuber yield at the end of the growing season. The infrared reflectance of the treatments without nematicide treatment in rotation B was only slightly higher than that for both treatments in rotation A during most of the growing season, but the decrease at the end of the growing season was slower. The plots in rotation B with nematicide treatment had a relatively high infrared reflectance during most of the growing season.

For rotation C the nematicide effect was more or less negative throughout the season. This was not seen in the tuber yield at the end of the season, although the nematicide effect on the latter variable was only minor. The level of infrared reflectance was intermediate between the two treatments of rotation B; this ranking coincided with the ranking according to tuber yield.

For rotation D the nematicide effect was negative during the first period of the growing season. During the second period the decrease in infrared reflectance was similar for both treatments. This coincided with the negative nematicide effect on tuber yield for this rotation. During the first part of the growing season the infrared reflectances for rotation D without nematicide and rotation B with nematicide were similar. However, the decrease in infrared reflectance at the end of the growing season for the treatments of rotation D was much slower than that in the other rotations, and therefore rotation D had the highest tuber yield and displayed a negative nematicide effect.

Although the relationship between amount of foliage (above-ground) and tuber yield (underground) is not straightforward, for the field trial described it was reasonably possible to explain differences in tuber yield by differences in infrared reflectance measured during the growing season.

## 7.9 Summary

For both sowing dates of field trial 116 in 1982 the crops showed a positive nitrogen effect on LAI on most sampling dates. During the first half of the season, LAI was larger in the early-sown crop than in the late-sown crop because of this difference in sowing date. Because senescence started earlier in the early-sown crop, LAI was largest in the late-sown crop during the second half of the season. At the highest nitrogen levels of the early-sown crop, some effect of lodging was noticeable. The nitrogen effect on green and red reflectances was opposite to that on LAI, because of the large absorption of visible radiation by green vegetation. The same applied for these reflectances as a function of time. With complete soil cover the reflectance in the visible passbands was very low and there was no clear nitrogen effect. At that stage there were differences in reflectance between the treatments of the two sowing dates, because of a difference in leaf colour. The nitrogen effects on infrared reflectance resembled those on LAI, which also applied to the infrared reflectance and LAI as a function of time. Results for reflectance factors were more stable in time than those for LAI. Similar results were found for the other field trials studied.

Critical levels in testing for treatment effects were systematically smaller for the infrared reflectance than for the LAI in all field trials. This leads to the conclusion that treatment effects were ascertained with larger power by means of reflectance measurements than by means of measurements in the field. The coefficients of variation of residuals for dry matter weight and LAI, especially the latter, were much larger than those for the reflectance values. The measurements of the infrared reflectance had particularly small CV values.

Results in estimating soil cover were poor because the ratio of reflectances or difference between reflectances in visible passbands were inaccurately ascertained.

For estimating LAI the growing season was subdivided into two stages: vegetative and generative. First, the corrected infrared reflectance was calculated by taking the difference between infrared and red reflectance. Then the LAI was estimated from the corrected infrared reflectance. For this latter regression two parameters had to be estimated (equation 5.12), which are different in the two stages. For the vegetative and the generative stages the regression of LAI on corrected infrared reflectance was described reasonably in all field trials by using this equation. Both stages were analysed separately. At the vegetative stage, some of the curves for the trials with barley were significantly different, some were not. The curves for the trials with wheat at the vegetative stage were not significantly different. At the generative stage, considerable discrepancies occurred, since LAI had been measured by different persons with slightly different instruments resulting in such large systematic discrepancies that direct comparison between LAI values predicted by reflectance measurements and those measured in the conventional manner becomes highly suspect. The best way to obtain LAI at the moment is to ascertain regression curves of LAI on corrected infrared reflectance for each situation and not to use regression curves from previous seasons. An important conclusion is that leaf angle distribution did not vary to such an extent that it disturbed the regression.

By using the regression curves for estimating LAI in the various field trials it was concluded that treatment effects were ascertained with larger power and that CV values were smaller for the LAI estimated by means of reflectance measurements than for the LAI measured in the field.

Ascertaining the regression curves for the vegetative and generative stages in only a few additional plots (field trial 100 in 1983), instead of using the trial itself (116 in 1983) for ascertaining the regression curves, gave approximately the same results.

Results for two other crops, grass and potatoes, indicated that the same theory and methodology could be applied to crops other than cereals.

# 8 FINAL REMARKS AND RECOMMENDATIONS

This chapter begins with a discussion about the plant characteristic that is most useful in agronomy and that may be estimated by remote sensing, then some pros and cons are given for multispectral aerial photography (MSP) vis-a-vis conventional field sampling. Next, sensor, calibration and dependence on weather are discussed and some restrictions applied in this study are enumerated. The chapter concludes with recommendations for further research.

## 8.1 WHAT PLANT CHARACTERISTIC IS MOST USEFUL IN AGRONOMY?

One aspect that needs further study in the near future is whether one should try to estimate LAI by means of reflectance measurements or whether one should look for some other plant characteristic in order to obtain information about crop growth or condition. Remote measurements of reflectances do not concern leaves only. Stems and ears, for instance, will also contribute to the measured reflectance. It is true that in most crops the leaves contribute the most to the measured reflectance. But the agronomist is not primarily interested in the LAI. He is more interested in the photosynthetic activity or light interception and its result: the final yield of some plant part. For pinpointing the cause of differences in yield he may also need information about the crop during the growing season. Photosynthetic activity is difficult to measure in the laboratory and it takes too much time and effort to measure it for plots in a field trial. The same applies to light interception. Therefore the agronomist has usually resorted to the LAI (since photosynthesis takes place primarily in the green leaves), or to dry matter weight. Remote sensing techniques offer new sorts of information about crops, and therefore in future agronomists may decide to use new plant characteristics that are more important than those currently used today.

If one is interested in plant characteristics other than LAI, such as dry matter content, the content of some nutrient (chemical analysis) or the mutual proportions of different plant parts, none of which can be ascertained directly by means of remote sensing techniques, one has to take samples in the field. However, the variability per plant of, e.g., some chemical component, is much smaller than the variability of LAI or dry matter weight per plant (see section 7.7). It is possible to ascertain the content of such a component with great accuracy by harvesting only a few plants. However, if one wishes to ascertain the total amount of some nutrient in the plants, one has to incorporate total fresh or dry matter weight into the calculations. These characteristics show a much larger variability within a plot (from plant to plant). If total fresh or dry matter weight correlate well with LAI, they may be ascertained more accurately, even though more indirectly, by means of reflectance measurements.

The best procedure if one is interested in this more detailed information will be to estimate LAI and fresh or dry matter weight by means of reflectance measurements and subsequently to ascertain plant composition by harvesting a few plants at random.

## 8.2 Pros and cons of MSP compared with conventional field sampling

In the present study it was shown that reflectance values obtained by MSP improved the power of testing for treatment effects in field trials. These measurements do provide quantitative information obtained instantaneously. Moreover, the information is obtained in a non-destructive way and, very importantly, is objective. The speed of availability of results, the labour intensity, the calibration of MSP and its dependence on the weather need to be discussed further (next sections).

In this study the system for MSP was only tested for utility. The costs of MSP were not compared with those of conventional field sampling techniques, because insufficient information was available for this comparison. Firstly, the costs of MSP largely depend on the extent to which the equipment is used (because investment is high and exploitation costs are low). Secondly, it is difficult to express in cash terms the improvement in accuracy obtained by using MSP rather than conventional field sampling. Therefore, no comparison was attempted.

### *8.2.1 Speed of availability of results, and labour intensity*

Multispectral aerial photography not only improves the power and accuracy of the testing for treatment effects (chapter 7), it can also be quicker to implement and requires less labour than conventional field sampling techniques.

With conventional field sampling techniques (chapter 6), plants have to be harvested in every plot. This is very labour intensive. The plant material then has to be analysed as soon as possible in the laboratory, because it is living. This leads to very large peaks of labour. Dry matter weight rather than fresh weight is usually ascertained, because field sampling takes time (at least several hours) and plants cut first often cannot be weighed immediately and will lose more water than the freshly cut plants: thus a comparison based on fresh weight is very unreliable.

Harvesting, weighing, ascertaining LAI, drying and reweighing the samples takes much time. If many field trials or a single large trial are to be analysed, sufficient assistance from experienced persons must be available, because all the operations are so labour intensive. In general, a researcher is only able to handle a very small number of field trials of restricted size. Nevertheless, it is a few days up to about a week before the results are available.

For obtaining aerial photographs (chapter 3) the labour intensity is lower. Within a few hours, many field trials can be recorded. For recording, a pilot, navigator and photographer are needed. One person can develop the film and analyse the images. A few days after recording, the results can be available; this means that the results are available as quickly as those from conventional field sampling.

Another aspect is that the chances of making mistakes with conventional measurements, e.g. by mixing up samples or by writing down the wrong weight readings, are large, unless all instruments are directly connected to a computer (which mostly will not be the case).

When measuring densities, all data are directly stored in a computer, which can immediately analyse them. This reduces the chance of introducing errors. On the other hand, it is also possible to postpone the analysis of the photographs to a much later date, if one is not immediately interested in the results. This means that the only time needed immediately is for obtaining the recordings, and maybe for developing the film; the other operations can be postponed until a more suitable time.

*8.2.2 Remarks about sensor and calibration*

Reflectance measurements cannot be analysed multitemporally unless the reflectances are well calibrated. It is not easy to calibrate a photographic system. Such a system first stores spectral information in analogue form and this subsequently has to be digitized in some way in order to obtain quantitative information. In the present study it was shown that it is possible to calibrate a photographic system satisfactorily in order to obtain calibrated reflectances relatively inexpensively (section 3.13). Aspects such as characteristic curve, light fall-off, lens and filter transmittance, relative aperture, exposure time and atmospheric influence were taken into account during the calibration. Most of these can be corrected for by using reference targets of known reflectance in the field. With multispectral aerial photography targets of about 5 m by 5 m will be usuable at a smallest scale of approximately 1:10 000 when the images are digitized by means of a densitometer with an aperture of 0.25 mm as used in this research. This will also be the minimum size of target that can be used with aerial multispectral scanning (spatial resolution worse than one square metre). However, in aerial photography, smaller targets can be used for measurements at scales larger than 1:10 000. For instance, at scale 1:3 000 the dimensions of the reference targets can be about 1.50 m by 1.50 m. Targets of such dimensions can remain in the field, if they are protected between missions.

It is not always possible or convenient to position reference targets near the recording site. Moreover, sometimes it is impossible to find a suitable location for setting up the targets. Another option may be the use of alternative targets that are permanently present in the area, such as roofs or roads. These will be almost constant over time in their reflectance characteristics. A body of water may be used in a similar way. The reflectances of such targets have to be measured once in the various passbands, e.g. with a radiometer. Homogeneous natural objects whose reflectance characteristics are not constant over time may also be used as calibration targets. For example, bare soil, a poorly developed crop and a well developed crop could be used. However, one has to take steps to avoid such targets introducing a random source of error. When using such objects the reflectance characteristics have to be measured with a radiometer during each mission, which costs additional effort. No such special prerequisites are needed for conventional field sampling.

### *8.2.3 Dependence on weather*

One of the main disadvantages of many remote sensing techniques in the optical region is their dependence on the weather. During recording, irradiation conditions have to be constant if a frame-to-frame comparison is to be valid. Also, wind and visibility may be limiting factors.

In this research we aimed for a recording frequency of once fortnightly during the growing season. This coincided with the sampling frequency applied in the field trials at the ir. A.P. Minderhoudhoeve, where this research was carried out. The 1982 and 1983 growing seasons indicated that it was possible to achieve a sampling frequency of two weeks.

The fact that information cannot be obtained at every desirable moment during the growing season can be a serious disadvantage of aerial photography compared with conventional techniques.

## 8.3 RESTRICTIONS APPLIED IN THE PRESENT STUDY

### *8.3.1 One crop*

In this monograph the data are almost all on cereals (more specifically, barley and wheat). In section 7.7 some results were given for grasses. Both crops belong to the family of Graminaceae, and are similar during the vegetative stage. Extrapolating our results to grasses will not meet great difficulties. Grasses are usually mown or grazed before they attain the generative stage. Thus, data on grass crops obtained by remote sensing do not have to be separated according to whether they relate to the vegetative or generative stage. In estimating a characteristic of grasses (e.g. dry matter weight) by remote sensing it may be necessary to incorporate a correction for the stubble that is left after mowing or grazing.

In principle, no great problems are anticipated in extrapolating the methodology tested in this research to other arable crops such as potatoes (see section 7.8) or sugar beet. In the same way as described in this monograph is should be possible to correct for background and to use remote sensing to estimate LAI and (above-ground) dry matter weight indirectly for these crops. It is merely a question of using other regression curves. However, a problem may arise if the agronomist is mainly interested in information about the plant part to be harvested at the end of the season. With potatoes and sugar beet these parts are in the soil. Often there is a clear correlation between a characteristic that can be estimated by remote sensing and one that the agronomist is interested in, e.g. LAI and the underground plant part, but this is by no means universal.

Remote sensing provides the agronomist with primary information about above-ground crop characteristics, but he has to know how to transform this information into information about underground plant parts.

More research should be focussed on ascertaining regression curves of LAI on corrected infrared reflectance for crops other than those investigated in this research. The relationship between LAI and dry matter weight (or fresh weight) for a vegetative crop means that regression curves of dry matter weight on corrected infrared reflectance must also be ascertained.

#### *8.3.2 One soil type*

In this monograph a procedure for correcting reflectances for soil background if soil moisture content or soil reflectance are not known explicitly was described (sections 5.5 and 5.6). This procedure requires soil types whose reflectance in any two passbands is linearly related, independent of soil moisture content. Although this will be a good approximation for many soil types, it will not be universal. For soil types whose reflectance in any two passbands is not linearly related, one has to investigate whether there is any relationship between reflectances in two passbands that is independent of soil moisture content. Otherwise, soil reflectances in various passbands have to be measured individually if the soil moisture content varies considerably during the investigation.

#### *8.3.3 No weeds*

One phenomenon that was not mentioned earlier, but which may disturb the relationship between measured reflectance and some crop characteristic, is the occurrence of weeds. In the field trials analysed in this study there were few weeds, so their influence could be ignored. However, in some cases weeds may be abundant. This not only applies to field trials, but even more to larger areas under normal farm management. Weeds may be so abundant that they contribute significantly to the measured reflectance. It will be very difficult to separate the contribution resulting from weeds from that resulting from the arable crops. Weeds may considerably disrupt the expected relationship between reflectance and some crop characteristic. For instance, at the end of the growing season more green weeds may occur in a badly developed wheat crop (e.g. low nitrogen level) than in a well developed crop (high nitrogen level). This may result in a higher infrared reflectance for the badly developed crop, which is opposite to what one would expect if no weeds were present.

### 8.4 RECOMMENDATIONS

Some ways in which the system could be improved and points for further research are given in this section.

1. The configuration of densitometer, digitizer and computer (described in section 3.6) was a prototype, which could be improved in the future. For example, the system could be made more fully automatized by driving the stepping

motors directly by a computer, allowing for more rapid positioning and a faster measuring speed.

2. If many field trials have to be recorded frequently during the growing season, the aeroplane should be quickly available at the recording site. For instance, at the experimental farm of the Wageningen Agricultural University an aeroplane should be stationed permanently in order to be flexible in fulfilling requests for aerial photographic recordings of field trials made by research workers. The aeroplane could be a microlight (see section 3.2) and personnel from the farm could fly it.
3. The positioning of reference targets near the recording site before each mission takes time and labour, and they cannot be set up at every location. Therefore, the possibilities of using objects such as roofs, roads, water bodies and vegetation as reference have to be studied. With low altitude aerial photography the possibilities of using small reference targets, which remain in the field between missions, should be investigated.
4. An important aspect that needs further study is which is the best plant characteristic on which information should be obtained by remote sensing. In this research LAI was used as an intermediate characteristic, because it could be measured conventionally as well as estimated via remote sensing, and it provides information about photosynthetic activity. In future, emphasis should be put on the possibilities of estimating plant characteristics, e.g. photosynthetic activity or light interception, directly from reflectances and on the possibility of using reflectances as plant characteristics in their own right.
5. The possibilities of extrapolating the results of the present study to other crops, other soil types and other plant parts also need to be studied further. Crop characteristics that correlate strongly with LAI may, in particular, yield good results. Also, possibilities of disease detection (in an early stage) with remote sensing need to be studied (see section 4.4.3). Results given in chapter 7 show the potential of remote sensing.
6. The possibility of applying the regression curve of LAI on reflectance for a specific crop or cultivar to other seasons and to other field trials with the same crop or cultivar, also has to be investigated further. The LAI measurements with the conventional sampling techniques in the present study showed large systematic discrepancies between field trials, due to the partly subjective measurement procedure, and therefore there is insufficient evidence to conclude that the regression curves of different crops or cultivars can be interchanged easily or applied in subsequent seasons.
7. Finally, in this research some aspects whose influence was deemed to be minor, were ignored. One such aspect is the adjacency effect. This effect may occur if two objects with distinctively different reflectances in a certain passband are adjacent. Then, an amount of radiant energy from the object with a high reflectance may be scattered by the atmosphere in the direction of the sensor and may finally be sensed as if it came from the object with a low reflectance.

Another aspect that has not been taken into account is angle-dependent reflectance. The view angle was always very close to the vertical downward (nadir). The recording was carried out as much as possible at about the same time of the day in order to minimize influences of the position of the sun.

By comparing reflectances ascertained by means of aerial photography with those obtained by means of radiometers, and by comparing the reflectances obtained by means of aerial photography in a multitemporal way, no large discrepancies were found that could be attributed to adjacency effects or angle-dependent reflectance.

Still, these aspects will need further attention and study in future research programmes.

# 9 MAIN CONCLUSIONS

1. Black and white multispectral aerial photography (MSP) was found to be the most appropriate technique for the frequent acquirement of remotely sensed data in field trials at the ir. A.P. Minderhoudhoeve experimental farm of the Wageningen Agricultural University, inasmuch as it offered characteristic spectral information about vegetation in the visible and infrared regions of the electromagnetic spectrum (chapters 2 and 3). By using cameras with 70-mm aerial films, mounted in an aeroplane, it was possible to obtain recordings at frequent intervals (high temporal resolution) cheaply. Film and filters were chosen carefully so that a high spectral resolution could be obtained (25-60 nm). The spatial resolution of the photographic material was sufficient for this application; spatial resolution was primarily limited by the densitometric processing system (conclusion 2). A complete field trial could be recorded instantaneously with one exposure.
2. By using a densitometer with an adequate aperture (0.25 mm) for obtaining quantitative measurements, a high spatial resolution was obtained (spot of 2 metres diameter at a recording scale of 1:8000). By interfacing this densitometer and a digitizing tablet with a computer, the film frame could be located easily with respect to the aperture of the densitometer (0.05 mm accuracy), and the recordings could be analysed quickly (1-2 hours per frame with the prototype used).
3. Reflectances in the green, red and infrared region of the electromagnetic spectrum ascertained with black and white multispectral aerial photography were well calibrated (section 3.13), offering specific information about an object (e.g. the crop). In this way quantitative information about extensive field trials was obtained instantaneously and non-destructively.
4. Information about crop reflectance obtained from the literature suggested that in the present study, reflectances in the visible region would be most suitable for estimating soil cover. For estimating LAI the reflectance in an infrared passband was expected to be more suitable than the reflectance in a visible passband because the penetration of infrared radiation into vegetation is larger than that of visible radiation (chapter 4).
5. If soil cover is redefined as the vertical projection of green vegetation and the relative area of the shadows included, perceived by a sensor pointing vertically downwards, relative to the total area, then the reflectance in a visible passband decreases linearly with increasing soil cover (section 5.2). Although a combination of green and red reflectances (section 5.5) could, theoretically, be used for correcting for the influence of soil background in estimating soil cover (a characteristic that is only of minor importance compared with LAI), results were poor because of the small differences between green and red reflectances. This conclusion is of no direct importance to the next conclusion.

6. It was shown to be possible to get around the problem of an unknown soil moisture content (and so an unknown soil reflectance) in estimating LAI (section 5.6), and thereby to obtain good results for the field trials analysed in the present study (chapter 7). Under the assumption that there was a constant ratio between the reflectance factors of bare soil in different passbands, independent of soil moisture content, a combination of green, red and infrared was applied for correcting the infrared reflectance for soil background (equation 5.25). This assumption holds for many soil types. For the special case that soil reflectance in the red and infrared were in the same order of magnitude, the infrared reflectance was corrected for soil background by calculating the difference between infrared and red reflectances (equation 5.28).
7. At the vegetative stage of cereals, the inverse of a special case of the Mitscherlich function, namely the one passing the origin (equation 5.12), was used for describing the regression function of LAI on the infrared reflectance corrected for background. This corrected infrared reflectance increases with increasing LAI, up to large LAI values (LAI 6-8, cf. conclusion 10).
8. A similar approach as in conclusion 7 was applied at the generative stage of cereals, by correcting the infrared reflectance for the background of yellow and dead leaves in ascertaining the difference between infrared and red reflectances (section 5.7).
9. When reflectance values obtained by MSP were compared with field measurements, in particular with LAI, information about treatment effects was acquired with larger power and with greater accuracy (smaller coefficients of variation = CV). Results for reflectances were also more stable in time than those for LAI (chapter 7). The reflectance values in an infrared passband were particularly outstanding (CV resulting from an analysis of variance for infrared reflectance ranged from 0.02-0.10 and CV for measured LAI ranged from 0.15-1.30 in the field trials analysed).
10. If LAI was estimated by reflectance values, by using a regression curve of LAI on corrected infrared reflectance, the critical levels in testing for treatment differences were in general smaller than for the measured LAI of samples. This also applied to the coefficients of variation (CV for estimated LAI ranged from 0.04-0.40 in the field trials analysed). Even at large LAI values (LAI 6-8) significant treatment effects could be distinguished by means of multispectral aerial photography, showing that the radiometric resolution of MSP was sufficient (relative radiometric resolution of 1.2% to about 2.2%).
11. In one field trial, the regression function of LAI on corrected infrared reflectance was ascertained by analysing a few additional plots (a training set). For these plots both LAI and reflectances were ascertained. The results show that to ascertain LAI in practice, no samples need to be gathered in the original field trial; this means that plots can be smaller than in the conventional situation, in which part of all the plots is used for sampling. To date there is insufficient evidence that the regression curves of different crops

or cultivars are easily transferable, or that the curve of one growing season can be applied in the following season, although the results pointed in that direction. This means that remote sensing can improve conventional field measurements, but cannot replace them. Conventional field measurements are still needed.

12. The LAI measured in the conventional manner appears to be a most unsatisfactory crop characteristic, because it is so hard to ascertain it objectively and accurately. Photosynthetic activity, for instance, may be better estimated by reflectance values. As a result, the application of remote sensing techniques may require a new definition of crop characteristics.

# SUMMARY


Remote sensing techniques enable quantitative information about a field trial to be obtained instantaneously and non-destructively. The aim of this study was to identify a method that can reduce inaccuracies in field trial analysis, and to identify how remote sensing can support and/or replace conventional field measurements in field trials.

In the literature there is a certain consensus that the best bands from which characteristic spectral information about vegetation can be extracted are those in the visible (green and red) and infrared regions of the electromagnetic spectrum. This was confirmed in the present study by an analysis of multispectral scanner data ('Daedalus scanner') from field trials with cereals. The optimal bands that were thereby selected for explaining grain yield mostly contained the channels 5 (550-600 nm), 7 (650-700 nm) and 9 (800-890 nm).

Multispectral aerial photography was found to be most appropriate for recording extensive field trials in a short period. In the present study, recordings were carried out with a single-engine aircraft, using two Hasselblad cameras for obtaining vertical photographs on black and white 70-mm aerial films. In this way, costs stayed within acceptable limits. The recording scale chosen, given the dimensions of the trials at the experimental farm of the Wageningen Agricultural University, where the research was carried out, was 1:8 000. Photographs were taken approximately fortnightly to keep in step with conventional field sampling. The film/filter combinations selected for obtaining a high spectral resolution and for matching bands 5, 7 and 9 of the Daedalus scanner, resulted in the following passbands:

green : 555-580 nm;
red : 665-700 nm;
infrared : 840-900 nm.

The densities of the objects on the film were measured by means of an automated Macbeth TD-504 densitometer. An aperture with a diameter of 0.25 mm was selected for the densitometer, in order to obtain a high spatial resolution at the scale of 1:8000, applicable to field trials with plots 3 metres wide. The measured densities were converted into exposure values, corrected for light fall-off, and then a linear function was applied to convert them into reflectance factors. In this linear function the exposure time, relative aperture, transmittance of the optical system, irradiance, path radiance and atmospheric attenuation were incorporated. Reference targets with known reflectance characteristics were set up in the field during missions and recorded at the same camera setting and under the same atmospheric conditions as the field trials, in order to ascertain the parameters of the linear function.

Information about crop reflectance obtained from the literature suggested that reflectances in the visible region of the electromagnetic spectrum (green or red) would be most suitable for estimating soil cover, whereas reflectances in the infrared might be most suitable for estimating leaf area index (LAI). Other plant characteristics, such as dry matter weight or yield, may be estimated indirectly from reflectances. Field trials with cereals analysed during the present study showed that treatment effects shown by green and red reflectances tended to be opposite to those shown by LAI. Treatment effects shown by infrared reflectance tended to be similar to those shown by LAI, even at large LAI (6-8). The treatment effects manifest in reflectances were more stable in time than those for LAI. Coefficients of variation of residuals resulting from analyses of variance were systematically smaller for reflectances than for the LAI in all experiments: those for the infrared reflectance were particularly small. In general, critical levels in testing for treatment effects were smaller for the infrared reflectance than for the LAI, which indicates that the power for infrared reflectance was larger than for LAI.

Soil moisture content is not constant during the growing season and differences in soil moisture content greatly influence soil reflectance. Since a multitemporal analysis of remote sensing data was required, a correction had to be made for soil background when ascertaining the relationship between reflectances and crop characteristics. In the literature no index or reflectance model stood out as being suitable for estimating crop characteristics in agricultural field trials. Thus, in this monograph an appropriate simplified reflectance model is presented for estimating soil cover and LAI for green vegetation. First of all, soil cover is redefined as: the vertical projection of green vegetation and the relative area of shadows included, seen by a sensor pointing vertically downwards, relative to the total soil area (in this definition soil cover depends on the position of the sun). Then, the simplified reflectance model is based on the expression of the measured reflectance as a composite reflectance of plants and soil: the measured reflectance in the various passbands is a linear combination of soil cover and its complement, with the reflectances of the plants and of the soil as coefficients, respectively.

By using this model, it should, theoretically, be possible to correct for soil background when estimating soil cover by combination of measurements in the green and red passbands. In practice, however, all the procedures derived yielded poor results because the difference between green and red reflectances was so small. Thus, attention was focussed on estimating LAI.

For estimating LAI a corrected infrared reflectance was calculated by subtracting the contribution of the soil from the measured reflectance. Theoretically, combining the reflectance measurements obtained in the green, red and infrared passbands, enables the corrected infrared reflectance to be calculated, without knowing soil reflectances. The main assumption was that there is a constant ratio between the reflectances of bare soil in different passbands, independent of soil moisture content: this assumption is valid for many soil types. For the soil type at the experimental farm of the Agricultural University, the corrected infrared reflectance can be approximated by the difference between total measured infrared and red reflectances. Subsequently this corrected infrared reflectance was used for estimating LAI according to the inverse of a special case of the Mitscherlich function. This function contains two parameters that have to be ascertained empirically. Model simulations with the SAIL model (introduced by Verhoef, 1984) confirmed the potential of this simplified, semi-empirical, reflectance model for estimating LAI.

Analogous derivations were applied for a generative canopy (cereals) with yellowing leaves.

The estimation of LAI by reflectances yielded good results for the field trials with cereals analysed in this study. The presence of treatment effects could be shown with larger power and the coefficients of variation were smaller for this estimated LAI than for the one measured in the field. Regression curves of LAI on corrected infrared reflectance differed significantly in different trials with the same crop, particularly for the generative stage. This may have been caused by large systematic discrepancies between LAI measurements obtained with the conventional sampling techniques for two field trials, because of subjectivity in separating green from yellow leaves. To date, the best approach is to ascertain regression curves of LAI on corrected infrared reflectance for each field trial by incorporating a few additional plots, in which both the LAI and the reflectances are measured.

# SAMENVATTING

Teledetectie-technieken bieden de mogelijkheid om op non-destructieve wijze informatie over een proefveld op één moment te verkrijgen. Doel van dit onderzoek was een methode te vinden om onnauwkeurigheden in de analyse van proefvelden te reduceren, en te bepalen op welke wijze teledetectie de gebruikelijke veldmetingen bij proefvelden kan ondersteunen en/of vervangen.

In de literatuur is men in het algemeen er over eens dat de beste banden om karakteristieke spectrale informatie over gewassen te verkrijgen, die in het zichtbare (groen en rood) en infrarode deel van het electromagnetische spectrum zijn. In dit onderzoek werd dit bevestigd door middel van een analyse van multispectrale scanner data ('Daedalus scanner') verkregen bij proefvelden met granen. De optimale banden, die daarbij geselecteerd werden om verschillen in korrelopbrengst te verklaren, bestonden meestal uit de banden 5 (550-600 nm), 7 (650-700 nm) en 9 (800-890 nm).

Voor de registratie van uitgebreide proefvelden in een kort tijdsbestek bleek multispectrale luchtfotografie het meest geschikt te zijn. In het huidige onderzoek werden opnamen verkregen met een eenmotorig vliegtuig, uitgerust met twee Hasselblad cameras om verticaal foto's te verkrijgen op zwart/wit 70-mm films. Op deze wijze bleven de kosten binnen de perken. Een opnameschaal van 1:8 000 werd toegepast vanwege de afmetingen van de proefvelden op het proefbedrijf van de Landbouwhogeschool, waar het onderzoek werd uitgevoerd. Opnamen werden ongeveer eens per twee weken gemaakt om gelijke tred te houden met de gebruikelijke monstername in het veld. De film/filter combinaties gekozen om een hoge spectrale resolutie te krijgen en om zoveel mogelijk samen te vallen met de banden 5, 7 en 9 van de Daedalus scanner, resulteerden in de volgende banden:

groen : 555-580 nm;
rood : 665-700 nm;
infrarood : 840-900 nm.

Densiteiten van de objecten werden op de film gemeten door middel van een geautomatiseerde Macbeth TD-504 densitometer. Voor de densitometer is een meetopening van 0.25 mm gekozen om een hoge spatiële resolutie te krijgen, die geschikt is voor proefvelden met plots van 3 meter breedte bij een opnameschaal van 1:8 000. De gemeten densiteiten werden herleid tot belichtingswaarden, gecorrigeerd voor lichtafval, en vervolgens met behulp van een lineaire functie herleid tot reflectiepercentages. In deze lineaire functie werden de belichtingstijd, effectieve lensopening, transmissie van het optische systeem, instraling, verstrooiing en absorptie in de atmosfeer betrokken. Ter bepaling van de parameters van de lineaire functie werden tijdens vluchten referentiepanelen met bekende reflectie-karakteristieken in het veld opgesteld en geregistreerd met dezelfde camera-instelling en onder dezelfde atmosferische omstandigheden als de proefvelden.

Informatie uit de literatuur over gewasreflectie gaf aan dat reflectiepercentages in het zichtbare deel van het electromagnetische spectrum (groen of rood) het meest geschikt zouden zijn om de bedekkingsgraad te schatten, terwijl reflectiepercentages in het infrarood het meest geschikt zouden zijn om de 'leaf area index' (LAI) te schatten. Andere kenmerken van planten, zoals droge-stofgewicht of opbrengst, zouden op indirecte wijze geschat kunnen worden uit reflectiepercentages. Proefvelden met granen, die tijdens dit onderzoek onderzocht zijn, hebben aangetoond dat behandelingseffecten voor de groen- en rood-reflectie een tendens hadden tegengesteld aan die voor de LAI. Behandelingseffecten voor de infrarood-reflectie leken op die voor de LAI, zelfs bij grote LAI (6-8). Behandelingseffecten waren bij de reflectiepercentages stabieler in de tijd dan bij de LAI. Variatiecoëfficiënten van residuen, resulterend uit variantie-analyses, waren voor de reflectiepercentages systematisch kleiner dan die voor de LAI bij alle onderzochte proefvelden: vooral die voor de infrarood-reflectie waren klein. Ook de overschrijdingskansen bij toetsing op behandelingseffecten waren over het algemeen voor de infrarood-reflectie kleiner dan die voor de LAI, hetgeen er op wijst dat met infrarood-reflectie een groter onderscheidingsvermogen werd bereikt dan met LAI.

Gedurende het groeiseizoen is het vochtgehalte van de bodem niet constant en verschillen in vochtgehalte beïnvloeden het reflectiepercentage van de bodem sterk. Aangezien een multitemporele analyse van de teledetectie-data vereist was, moest er een correctie voor de bodem toegepast worden om de relatie tussen reflectie en gewaskenmerken te bepalen. In de literatuur werd geen index of reflectiemodel gevonden dat erg geschikt was om gewaskenmerken te schatten bij proefvelden. Daarom is er in deze monografie een geschikt, vereenvoudigd reflectiemodel ingevoerd om de bedekkingsgraad en de LAI voor groene vegetatie te schatten. Allereerst is de definitie van bedekkingsgraad herzien tot: de verticale projectie van groene vegetatie met het relatieve oppervlak van de schaduw inbegrepen, zichtbaar voor een sensor die verticaal naar beneden gericht is, relatief ten opzichte van het totale bodemoppervlak (in deze definitie hangt de bedekkingsgraad af van de positie van de zon). Met deze nieuwe definitie voor bedekkingsgraad, is het vereenvoudigde reflectiemodel gebaseerd op het uitdrukken van de gemeten reflectie als de samengestelde reflectie van planten en bodem: het gemeten reflectiepercentage in de verschillende banden is een lineaire combinatie van de bedekkingsgraad en zijn complement met als respectievelijke coëfficiënten de reflectiepercentages van plant en bodem.

Met gebruik van dit model zou het theoretisch mogelijk zijn een correctie voor de bodem toe te passen bij de schatting van de bedekkingsgraad door het combineren van metingen in het groen en rood. Alle procedures, die afgeleid zijn, gaven in de praktijk vrij slechte resultaten ten gevolge van het kleine verschil tussen groen- en rood-reflectie. Daarom werd de aandacht gericht op schatting van de LAI.

Om de LAI te schatten werd een gecorrigeerde infrarood-reflectie berekend door de bijdrage van de bodem in mindering te brengen op het gemeten reflectiepercentage. Door het combineren van reflectiemetingen verkregen in de groene, rode en infrarode banden is het theoretisch mogelijk de gecorrigeerde infrarood-reflectie te bepalen, zonder de reflectie van de bodem te kennen. De belangrijkste veronderstelling was dat er een constante verhouding bestaat tussen de reflectiepercentages van kale bodem in verschillende banden, onafhankelijk van het vochtgehalte van de bodem (hetgeen voor veel bodemsoorten geldt). Voor de bodemsoort op het proefbedrijf van de Landbouwhogeschool kan men de gecorrigeerde infrarood-reflectie benaderen door middel van het verschil tussen totaal gemeten infrarood- en rood-reflectie. Vervolgens werd de gecorrigeerde infrarood-reflectie gebruikt om de LAI te schatten volgens de inverse van een bijzonder geval van de Mitscherlich functie. Deze functie bevat twee parameters die empirisch bepaald moeten worden. Model-simulatie met het SAIL-model (geïntroduceerd door Verhoef, 1984) bevestigden de mogelijkheden van dit vereenvoudigd, semi-empirisch, reflectiemodel om de LAI te schatten.

Voor het generatieve gewas (granen) met vergelende bladeren zijn analoge afleidingen toegepast.

Voor de in dit onderzoek geanalyseerde proefvelden met granen gaf de schatting van de LAI door middel van reflectiepercentages goede resultaten. De aanwezigheid van behandelingseffecten kon met een groter onderscheidingsvermogen aangetoond worden en de variatiecoëfficiënten waren kleiner voor de aldus geschatte LAI dan voor die gemeten in het veld. Regressiecurven van LAI op gecorrigeerde infrarood-reflectie waren significant verschillend in verschillende proeven met één en hetzelfde gewas, vooral voor het generatieve gewas. Dit kan het gevolg geweest zijn van systematische verschillen tussen metingen van de LAI volgens de gebruikelijke methode van monstername bij twee veldproeven, vanwege subjectiviteit bij de scheiding in groene en gele bladeren. Op dit moment is de beste procedure om regressiecurven van LAI op gecorrigeerde infrarood-reflectie voor elk proefveld te bepalen door het opnemen van enkele extra veldjes, waarbij zowel de LAI als ook de reflectiepercentages gemeten worden.

# REFERENCES


AASE, J.K., F.H. SIDDOWAY & J.P. MILLARD, 1984. Spring wheat-leaf phytomass and yield estimates from airborne scanner and hand-held radiometer measurements. Int. J. Rem. Sens. **5:** 771-781.

AHLRICHS, J.S. & M.E. BAUER, 1983. Relation of agronomic and multispectral reflectance characteristics of spring wheat canopies. Agron. J. **75:** 987-993.

ALLEN, W.A., T.V. GAYLE & A.J. RICHARDSON, 1969. Interaction of light with a compact leaf. J. Opt. Soc. Am. **59:** 1376-1379.

ALLEN, W.A. & A.J. RICHARDSON, 1968. Interaction of light with a plant canopy. J. Opt. Soc. Am. **58:** 1023-1028.

ASRAR, G., M. FUCHS, E.T. KANEMASU & J.L. HATFIELD, 1984. Estimating absorbed photosynthetic radiation and leaf area index from spectral reflectance in wheat. Agron. J. **76:** 300-306.

BADHWAR, G.D., 1980. Crop emergence date determination from spectral data. Photogram. Eng. Rem. Sens. **46:** 369-377.

BARNETT, T.L. & D.R. THOMPSON, 1983. Large-area relation of Landsat MSS and NOAA-6 AVHRR spectral data to wheat yields. Rem. Sens. Envir. **13:** 277-290.

BARNSLEY, M.L., 1984. Integration of multispectral data obtained at different view angles for vegetation analysis. Proc. EARSeL/ESA Symp. on Integrated Approaches in Remote Sensing, Guildford, U.K., ESA SP-214, p. 135-142.

BAUER, M.E., L.L. BIEHL & B.F. ROBINSON, 1980. Field research on the spectral properties of crops and soils. AgRISTARS, Final Report, November 1980, Vol. 1, Purdue Univ., W-Lafayette, SR-PO-0422; NAS9-15466.

BAUER, M.E., and STAFF, 1982. Remote sensing of agricultural crops and soils. AgRISTARS, Final Report, Purdue Univ., W-Lafayette, SR-P2-04266; NAS9-15466.

BOWERS, S.A. & R.J. HANKS, 1965. Reflection of radiant energy from soils. Soil Science **100:** 130-138.

BROWN, R.J. & F.J. AHERN, 1980. The field spectral measurements program of the Canada centre for remote sensing. Can. J. Rem. Sens. **6:** 26-37.

BUNNIK, N.J.J., 1978. The multispectral reflectance of shortwave radiation by agricultural crops in relation with their morphological and optical properties. Thesis, Meded. Landbouwhogeschool Wageningen 78-1, 175 pp.

BUNNIK, N.J.J., 1981. Fundamentals of remote sensing. Relations between spectral signatures and physical properties of crops. In: Application of Remote Sensing to Agricultural Production Forecasting. ed. A. Berg; A.A. Balkema, Rotterdam, p. 47-81.

BUNNIK, N.J.J., 1984. Review of models and measurements of multispectral reflectance by plant canopies. Recommendations for future research. Publ. NLR MP 84039 U, 11 pp.

BUNNIK, N.J.J. & W. VERHOEF, 1974. The spectral directional reflectance of agricultural crops. Measurements on a wheat and a grass canopy for some stages of growth. Niwars publ. **23:** 123 pp.

BUNNIK, N.J.J. AND STAFF, 1977. Onderzoek naar de toepassingsmogelijkheden van multispectrale scanning. Niwars publ. **44:** 371 pp.

CHANCE, J.E., 1977. Applications of Suits spectral model to wheat. Rem. Sens. Envir. **6:** 147-150.

CHANCE, J.E. & E.W. LEMASTER, 1977. Suits reflectance models for wheat and cotton: theoretical and experimental tests. Appl. Opt. **16:** 407-412.

COCHRAN, W.G. & G.M. COX, 1950. Experimental designs. John Wiley & Sons, Inc., New York.

COLWELL, J.E., 1974. Vegetation canopy reflectance. Rem. Sens. Envir. **3:** 175-183.

CONDIT, H.R., 1970. The spectral reflectance of American soils. Photogram. Eng. Rem. Sens. **36:** 955-966.

COX, D.R., 1958. Planning of experiments. John Wiley & Sons, Inc., New York.

CURRAN, P.J., 1980. Relative reflectance data from preprocessed multispectral photography Int. J. Rem. Sens. **1:** 77-83.

CURRAN, P.J., 1981. Remote sensing: The role of small format light aircraft photography. Geographical papers no 75, Dep. of Geography, Univ. of Reading, England, 39 pp.

Curran, P.J., 1982a. Multispectral photographic remote sensing of green vegetation biomass and productivity. Photogram.Eng. Rem. Sens. **48:** 243-250.
Curran, P.J., 1982b. Polarized visible light as an aid to vegetation classification. Rem. Sens. Envir. **12:** 491-499.
Curran, P.J., 1983. Estimating green LAI from multispectral aerial photography. Photogram. Engr. Rem. Sens. **49:** 1709-1720.
Daniel, C. & F.S. Wood, 1980. Fitting equations to data. John Wiley & Sons, Inc., New York.
Daughtry, C.S.T. & S.E. Hollinger, 1984. Costs of measuring leaf area index of corn. Agron. J. **76:** 836-841.
Duntley, S.Q., 1942. The optical properties of diffusing materials. J. Opt. Soc. Am. **32:** 61-70.
Egan, W.G., 1985. Photometry and polarization in remote sensing. Elsevier Science Publishing Co., Inc., New York.
Gausman, H.W., 1982. Visible light reflectance, transmittance, and absorptance of differently pigmented cotton leaves. Rem. Sens. Envir. **13:** 233-238.
Gausman, H.W., W.A. Allen, C.L. Wiegand, D.E. Escobar, R.R. Rodriguez & A.J. Richardson., 1973. The leaf mesophylls of twenty crops, their light spectra, and optical and geometrical parameters. U.S.D.A., Techn. Bull. no 1465, Washington D.C., 59 pp.
Goillot, Ch.C., 1980. Significance of spectral reflectance for natural surfaces. In: Remote Sensing Application in Agriculture and Hydrology. ed. G. Fraysse; A.A. Balkema, Rotterdam; pp. 53-68.
Goudriaan, J., 1977. Crop micrometeorology: a simulation study. Thesis Landbouwhogeschool, Wageningen, 249 pp.
Graham, R., 1980. The ITC multispectral camera system with respect to crop prognosis in winterwheat. ITC Journal 1980-2, 235-254.
Gray, T.I. Jr. & D.G. McCrary, 1981a. Meteorological satellite data – A tool to describe the health of the world's agriculture. AgRISTARS Report EW-N1-04042; JSC-17712, Johnson Space Flight Center, Houston, Texas.
Gray, T.I. Jr. & D.G. McCrary, 1981b. The environmental vegetation index: a tool potentially useful for arid land management. Report no. E81-10191; NASA-CR-160978, NOAA, Houston, Texas.
Hatfield, J.L., 1981. Spectral behavior of wheat yield variety trials. Photogram. Engr. Rem. Sens. **47:** 1487-1491.
Hatfield, J.L., 1983. Remote sensing estimators of potential and actual crop yield. Rem. Sens. Envir. **13:** 301-311.
Hatfield, J.L., G. Asrar & E.T. Kanemasu, 1984. Intercepted photosynthetically active radiation estimated by spectral reflectance. Rem. Sens. Envir. **14:** 65-75.
Hatfield J.L., J.P. Millard & R.C. Goettelman, 1982. Variability of surface temperature in agricultural fields of Central California. Photogram. Engr. Rem. Sens. **48:** 1319-1325.
Heilman, J.L. & D.G. Moore, 1982. Evaluating near-surface soil moisture using heat capacity mapping mission data. Rem. Sens. Envir. **12:** 117-121.
Hlavka, C.A., S.M. Carlyle, R. Yokoyama & R.M. Haralick, 1979. Multi-temporal classification of winter wheat using a growth state model. Proc. Symp. on Mach. Proc. Rem. Sens. Data, Purdue Univ., W-Lafayette, Ind., 105-115.
Holben, B.N., C.J. Tucker & C.J. Fan, 1980. Spectral assessment of soybean leaf area and leaf biomass. Photogram. Engr.Rem. Sens. **46:** 651-656.
Horler, D.N.H., M. Dockray & J. Barber, 1983. The red edge of plant leaf reflectance. Int. J. Rem. Sens. **4:** 273-288.
Idso, S.B. & C.T. De Wit, 1970. Light relations in plant canopies. Appl. Opt. **9:** 177-184.
Idso, S.B., R.J. Reginato, D.C. Reicosky & J.L. Hatfield, 1981. Determining soil-induced plant water potential depressions in Alfalfa by means of infrared thermometry. Agron. J. **73:** 826-830.
Jackson, R.D., R.J. Reginato, P.J. Pinter Jr. & S.B. Idso, 1979. Plant canopy information extraction from composite scene reflectance of row crops. Appl. Opt. **18:** 3775-3782.
Jackson, R.D., P.N. Slater & P.J. Pinter Jr., 1983a. Discrimination of growth and water stress in wheat by various vegetation indices through clear and turbid atmospheres. Rem. Sens. Envir. **13:** 187-208.

JACKSON, R.D., P.N. SLATER & P.J. PINTER JR., 1983b. Adjusting the tasselled-cap brightness and greenness factors for atmospheric path radiance and absorption on a pixel by pixel basis. Int. J. Rem. Sens. **4:** 313-323.

JANSE, A.R.P. & N.J.J. BUNNIK, 1974. Reflectiespectra van enige Nederlandse bodemmonsters bepaald met de NIWARS-veldspectrometer. NIWARS-publ. **18**: 21 pp.

JANSEN, A.A.M., 1978. Zoeken naar een geschikt regressiemodel. VVS Bull. **11:** 16-22.

JONKERS, G., 1982. Referentiepanelen voor waarnemingen vanuit de lucht. TNO Rapp. Nr. V-82-173, 46 pp.

KANNEGIETER, A., 1980. An experiment using multispectral photography for the detection and damage assessment of disease infection in winter-wheat: Agronomic considerations. ITC Journal 1980-2: 189-234.

KAUTH, R.J. & G.S. THOMAS, 1976. The tasseled cap – A graphic description of the spectral-temporal development of agricultural crops as seen by Landsat. Proc. Symp. on Mach. Proc. Rem. Sens. Data, Purdue Univ., W-Lafayette, Ind., **4B:** 41-51.

KIMES, D.S., 1983. Dynamics of directional reflectance factor distributions for vegetation canopies. Appl. Opt. **22:** 1364.

KIMES, D.S. & J.A. KIRCHNER, 1982. Radiative transfer model for heterogeneous 3-D scenes. Appl. Opt. **21:** 4119-4129.

KODAK FILTERS FOR SCIENTIFIC AND TECHNICAL USES. 1978. Kodak publication No. B-3, 92 pp.

KODAK DATA FOR AERIAL PHOTOGRAPHY. 1976. Kodak publication No. M-29, 92 pp.

KOEPKE, P. & K.T. KRIEBEL, 1978. Influence of measured reflection properties of vegetated surfaces on atmospheric radiance and its polarization. Appl. Opt. **17:** 260-264.

KONDRATYEV, K.YA. & O.M. POKROVSKY, 1979. A factor analysis approach to optimal selection of spectral intervals for multipurpose experiments in remote sensing of the environment and earth resources. Rem. Sens. Envir. **8:** 3-10.

KUBELKA, P. & F. MUNK, 1931. Ein Beitrag zur Optik der Farbanstriche. A. Techn. Physik **11:** 593-601.

LARGE, E.C., 1954. Growth stages in cereals. Illustration of the Feekes scale. Plant Pathol. **3:** 128-129.

LEMASTER, E.W., 1975. Further tests of plant canopy reflectance models and investigation of non-Lambertian properties of plant canopies. NASA report CR-145615.

LILLESAND, T.M. & R.W. KIEFER, 1979. Remote sensing and image interpretation. J. Wiley & Sons, Inc., New York, 612 pp.

MALILA, W.A., 1968. Multispectral techniques for image enhancement and discrimination. Photogram. Eng. **34:** 566-575.

THE MANUAL OF PHOTOGRAPHY, 1978. Seventh edition. Focal Press (Butterworth & Co. Ltd.), London, Boston.

MANUAL OF REMOTE SENSING, 1983. Second edition. (R.N. Colwell, ed.), American Society of Photogrammetry, Falls Church, Virginia.

MARKHAM, B.L., D.S. KIMES, C.J. TUCKER & J.E. MCMURTREY III, 1981. Temporal spectral response of a corn canopy. Photogram. Eng. Rem. Sens. **48:** 1599-1605.

MILLER, G.P., M. FUCHS, M.J. HALL, G. ASRAR, E.T. KANEMASU & D.E. JOHNSON, 1984. Analysis of seasonal multispectral reflectances of small grains. Rem. Sens. Envir. **14:** 153-167.

MILTON, E.J., 1980. A portable multiband radiometer for ground data collection in remote sensing. Int. J. Rem. Sens. **1:** 153-165.

OTT, W., B. PFEIFFER & F. QUIEL, 1984. Directional reflectance properties determined by analysis of airborne multispectral scanner data and atmospheric correction. Rem. Sens. Envir. **16:** 47-54.

PEARSON, R.L. & L.D. MILLER, 1971. Design of field spectrophotometer lab. Grassland Biome, Techn. Rep. 133:, 102 pp.

PEARSON, R.L. & L.D. MILLER, 1973a. Mapping standing crop biomass by computer processing of aerial multispectral imagery. Grassland Biome, Techn. Rep. 214, 32 pp.

PEARSON, R.L. & L.D. MILLER, 1973b. The biometer: A hand-held biomass meter. Grassland Biome, Techn. Rep. 221, 24 pp.

PEARSON, R.L., L.D. MILLER & C.J. TUCKER, 1976. Hand-held spectral radiometer to estimate gramineous biomass. Appl. Opt. **15:** 416-418.

PINTER, P.J. JR., R.D. JACKSON, S.B. IDSO & R.D. REGINATO, 1981. Multidate spectral reflectance as predictors of yield in water stressed wheat and barley. Int. J. Rem. Sens. **2:** 43-48.

POLLOCK, R.B. & E.T. KANEMASU, 1979. Estimating leaf-area index of wheat with Landsat data. Rem. Sens. Envir. **8:** 307-312.

RAO, V.R., E.J. BRACH & A.R. MACK, 1979. Bidirectional reflectance of crops and the soil contribution. Rem. Sens. Envir. **8:** 115-125.

RICHARDSON, A.J. & C.L. WIEGAND, 1977. Distinguishing vegetation from soil background information. Photogram. Eng. Rem. Sens. **43:** 1541-1552.

ROSS, D.S., 1973. Atmospheric effects in multispectral photographs. Photogram. Engr. **39:** 377-384.

ROUSE, J.W. JR., R.H. HAAS, J.A. SCHELL & D.W. DEERING, 1973. Monitoring vegetation systems in the Great Plains with ERTS. Earth Res. Techn. Satellite-1 Symp., Goddard Space Flight Center, Washington D.C., pp. 309-317.

ROUSE, J.W. JR., R.H. HAAS, D.W. DEERING, J.A. SCHELL & J.C. HARLAN, 1974. Monitoring the vernal advancement and retrogradation (green wave effect) of natural vegetation. NASA/GSFC Type III Final Report, Greenbelt, Md., 371 pp.

SCARPACE, F.L., 1978. Densitometry on multi-emulsion imagery. Photogram. Eng. Rem. Sens. **44:** 1279-1292.

SCARPACE, F.L. & G.A. FRIEDERICHS, 1978. A method of determining spectral analytical dye densities. Photogram. Eng. Rem. Sens. **44:** 1293-1301.

SCARPACE, F.L. & B.K. QUIRK, 1982. Applications of analytical densitometry to remote sensing. J. Appl. Photographic Eng. **8:** 147-152.

SCHNUTE, J., 1981. A versatile growth model with statistically stable parameters. Can. J. Fish. Aquat. Sci. **38:** 1128-1140.

SIEVERS, J., 1976. Zusammenhänge zwischen Objectreflexion und Bildschwärzung in Luftbildern. München, Bayerischen Akademie der Wissenschaften, Reihe C: Dissertationen, Heft Nr. 221, 129 pp.

SLATER, P.N., 1980. Remote sensing: optics and optical systems. Addison-Wesley Publ. Comp., Reading, Massachusetts.

SLATER, P.N. & R.D. JACKSON, 1982. Atmospheric effects on radiation reflected from soil and vegetation as measured by orbital sensors using various directions. Appl. Opt. **21:** 3923.

SNEDECOR G.W. & W.G. COCHRAN, 1980. Statistical Methods. Seventh edition. Iowa State University Press, Iowa.

SPIERTZ, J.H.J. & J. ELLEN, 1978. Effects of nitrogen on crop development and grain growth of winter wheat in relation to assimilation and utilization of assimilates and nutrients. Neth. J. Agric. Sci. **26:** 210-231.

STEVEN , M.D., P.V. BISCOE & K.W. JAGGARD, 1983. Estimation of sugar beet productivity from reflection in the red and infrared spectral bands. Int. J. Rem. Sens. **4:** 325-334.

STONER, E.R., M.F. BAUMGARDNER, L.L. BIEHL & B.F. ROBINSON, 1980. Atlas of soil reflectance properties. Agric. Exp. Station, Purdue Univ., W-Lafayette, Indiana, Res. Bull. 962, 75 pp.

SUITS, G.H., 1972a. The calculation of the directional reflectance of a vegetative canopy. Rem. Sens. Envir. **2:** 117-125.

SUITS, G.H., 1972b. The cause of azimuthal variations in directional reflectance of vegetative canopies. Rem. Sens. Envir. **2:** 175-182.

SUITS, G.H. & G.R. SAFIR, 1972. Verification of a reflectance model for mature corn with applications to corn blight detection. Rem. Sens. Envir. **2:** 183-192.

SWAIN, P.H. & S.M. DAVIS, 1978. Remote sensing: The quantitative approach. McGraw-Hill, New York.

THOMPSON, D.R. & O.A. WEHMANEN, 1980. Using Landsat digital data to detect moisture stress in corn-soybean growing regions. Photogram. Eng. Rem. Sens. **46:** 1087-1093.

TUCKER, C.J., 1979. Red and photographic infrared linear combinations for monitoring vegetation. Rem. Sens. Envir. **8:** 127-150.

TUCKER, C.J., 1980. A spectral method for determining the percentage of green herbage material in clipped samples. Rem. Sens. Envir. **9:** 175-181.

TUCKER, C.J., J.H. ELGIN JR., J.E. MCMURTREY III & C.J. FAN, 1979. Monitoring corn and soybean crop development with hand-held radiometer spectral data. Rem. Sens. Envir. **8:** 237-248.

TUCKER, C.J., J.U. HIELKEMA & J. ROFFEY, 1985. The potential of satellite remote sensing of ecological conditions for survey and forecasting desert-locust activity. Int. J. Rem. Sens. **6:** 127-138.

TUCKER, C.J., B.N. HOLBEN, J.H. ELGIN JR. & J.E. MCMURTREY, 1980. Relationship of spectral data to grain yield variation. Photogram. Eng. Rem. Sens. **46:** 657-666.

TUCKER, C.J., L.D. MILLER & R.L. PEARSON, 1973. Measurement of the combined effect of green biomass, chlorophyll, and leaf water on canopy spectroreflectance of the shortgrass prairie. Proc. of the annual Rem. Sens. Earth Res. Conf., 2nd, 1-27.

UENK, D., 1982. Bepaling van grondbedekking en biomassa met behulp van een reflectiemeter. CABO-Verslag nr. 41, 18 pp.

VERHOEF, W., 1979. An improved feature selection criterion and its application to multispectral reflectance data from crops. NLR, Amsterdam, Publ. NLR TR 79002 U, 66 pp.

VERHOEF, W., 1984. Light scattering by leaf layers with application to canopy reflectance modelling: The SAIL model. Rem. Sens. Envir. **16:** 125-141.

VERHOEF, W. & N.J.J. BUNNIK, 1974. Spectral reflectance measurements on agricultural field crops during the growing season. Niwars publ. **31**, 72 pp.

VERHOEF, W. & N.J.J. BUNNIK, 1975. A model study on the relations between crop characteristics and canopy spectral reflectance. Niwars publ. **33**, 89 pp.

VERHOEF, W. & N.J.J. BUNNIK, 1976. The spectral directional reflectance of row crops. Niwars publ. **35**, 134 pp.

VERHOEF, W. & N.J.J. BUNNIK, 1981. Influence of crop geometry on multispectral reflectance determined by the use of canopy reflectance models. Proc. Int. Coll. on Signatures of Remotely Sensed Objects, Avignon, France, pp. 273-290.

WARDLEY, N.W. & P.J. CURRAN, 1984. The estimation of green-leaf-area index from remotely sensed airborne multispectral scanner data. Int. J. Rem. Sens. **5:** 671-679.

YOUKHANA, S.K., 1983. Canopy modelling studies. Colorado State Univ., PhD., 84 pp.

APPENDIX 1 Growth stages of cereals on the Feekes scale (Large, 1954).

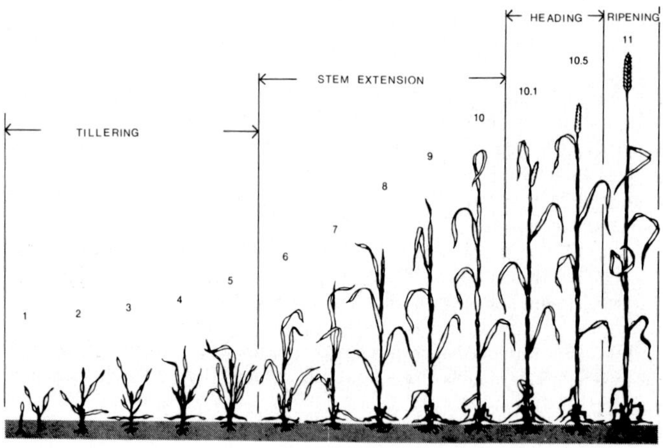


APPENDIX 2 Plan of the ir. A.P. Minderhoudhoeve, experimental farm of the Wageningen Agricultural University. Three flight lines are drawn across the region with arable crops.

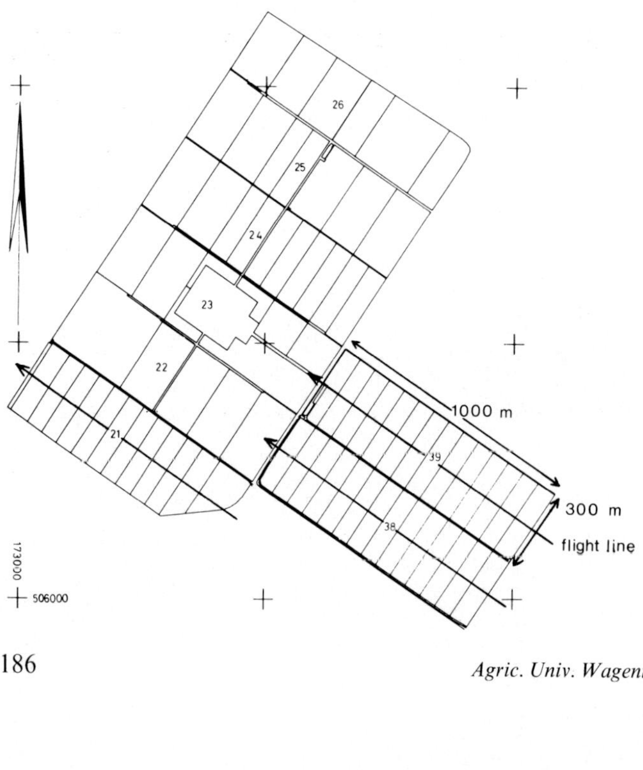

APPENDIX 3 Comparison of reflectance factors ascertained by means of multispectral photography (MSP) with those obtained by means of a hand-held radiometer (HRM) and a field spectroradiometer (FSM) in field trial 116 in 1983.

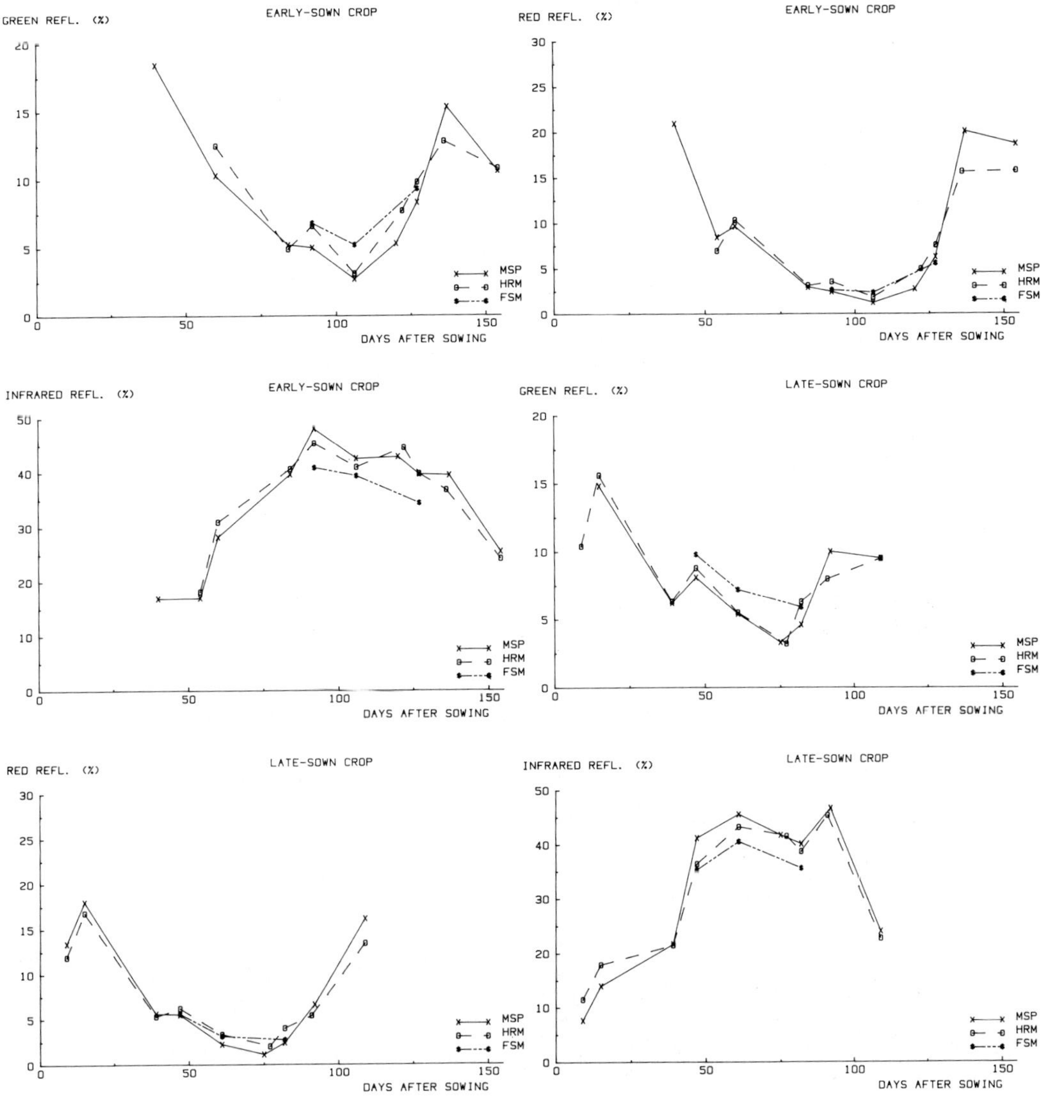

APPENDIX 4 Comparison of reflectance factors ascertained by means of multispectral photography (MSP) with those obtained by means of a hand-held radiometer (HRM) and a field spectroradiometer (FSM) in field trial 92 in 1983.

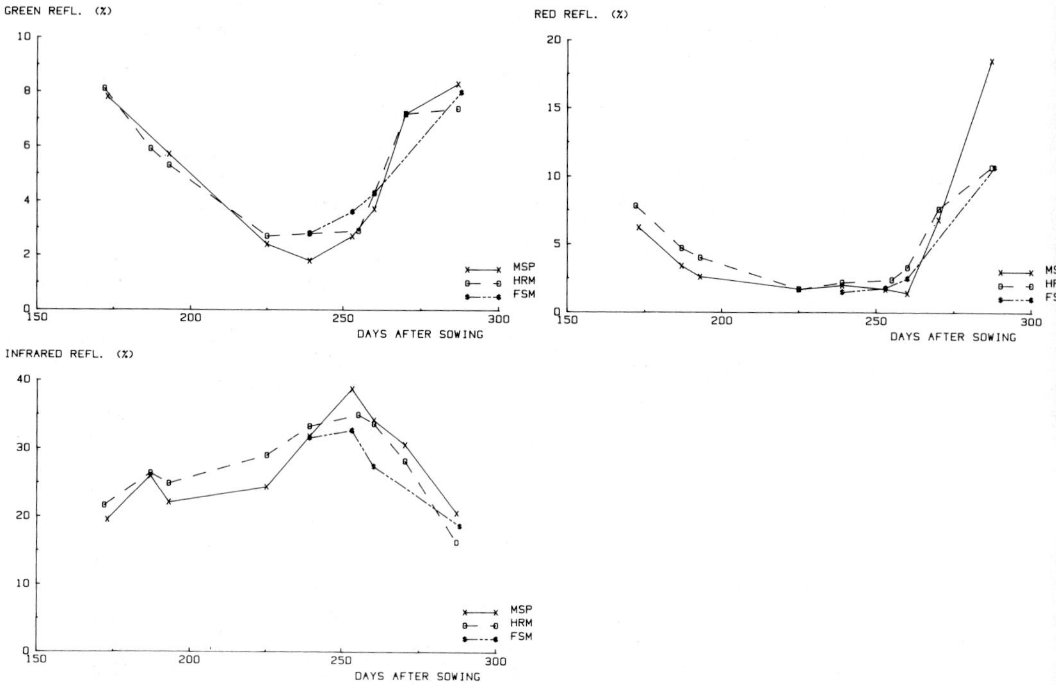


APPENDIX 5 Model simulations by means of the SAIL model (cf. section 5.8) for a spherical, planophile and erectophile leaf angle distribution. Irradiance conditions were either direct sunlight or diffuse skylight. Calculations were carried out for three soil conditions: dry, wet and black soil. In addition to green, red and infrared reflectance factors, soil cover with the traditional definition ('old') and with the new definition ('new') was also calculated. The data set was provided by Verhoef.

| leaf angle (degrees) | relative frequency | | |
|---|---|---|---|
| | spherical | planophile | erectophile |
| 5 | 1.5 | 22.0 | 0.2 |
| 15 | 4.5 | 20.7 | 1.5 |
| 25 | 7.4 | 18.2 | 4.0 |
| 35 | 10.0 | 14.9 | 7.3 |
| 45 | 12.3 | 11.1 | 11.1 |
| 55 | 14.3 | 7.3 | 15.9 |
| 65 | 15.8 | 4.0 | 18.2 |
| 75 | 16.8 | 1.5 | 20.7 |
| 81 | 3.4 | 0.1 | 4.3 |
| 83 | 3.5 | 0.1 | 4.4 |
| 85 | 3.5 | 0.0 | 4.4 |
| 87 | 3.5 | 0.0 | 4.4 |
| 89 | 3.5 | 0.0 | 4.4 |

1 Spherical leaf angle distribution; dry soil

| LAI | reflectance factor (%) | | | | | | soil cover (%) | |
|---|---|---|---|---|---|---|---|---|
| | sunlight | | | skylight | | | old def. | new def. |
| | G | R | IR | G | R | IR | | |
| 0.0 | 20.0 | 22.0 | 24.2 | 20.0 | 22.0 | 24.2 | 0.0 | 0.0 |
| 0.1 | 18.2 | 19.7 | 24.6 | 17.8 | 19.2 | 24.7 | 4.9 | 11.4 |
| 0.2 | 16.6 | 17.7 | 25.0 | 15.9 | 16.9 | 25.3 | 9.5 | 21.5 |
| 0.3 | 15.1 | 15.9 | 25.4 | 14.2 | 14.8 | 25.8 | 14.0 | 30.4 |
| 0.4 | 13.8 | 14.3 | 25.8 | 12.8 | 13.0 | 26.4 | 18.2 | 38.3 |
| 0.5 | 12.7 | 12.9 | 26.2 | 11.5 | 11.4 | 27.0 | 22.2 | 45.4 |
| 0.6 | 11.6 | 11.6 | 26.6 | 10.4 | 10.1 | 27.5 | 26.0 | 51.6 |
| 0.7 | 10.7 | 10.5 | 27.1 | 9.5 | 8.9 | 28.1 | 29.6 | 57.1 |
| 0.8 | 9.8 | 9.5 | 27.5 | 8.6 | 7.9 | 28.6 | 33.1 | 62.0 |
| 0.9 | 9.1 | 8.6 | 27.9 | 7.9 | 7.0 | 29.1 | 36.3 | 66.3 |
| 1.0 | 8.4 | 7.8 | 28.4 | 7.2 | 6.3 | 29.7 | 39.5 | 70.1 |
| 1.2 | 7.3 | 6.4 | 29.2 | 6.2 | 5.1 | 30.7 | 45.2 | 76.6 |
| 1.4 | 6.3 | 5.4 | 30.1 | 5.4 | 4.1 | 31.6 | 50.5 | 81.6 |
| 1.6 | 5.6 | 4.5 | 30.9 | 4.8 | 3.4 | 32.5 | 55.2 | 85.5 |
| 1.8 | 5.0 | 3.8 | 31.7 | 4.4 | 2.9 | 33.4 | 59.5 | 88.7 |
| 2.0 | 4.6 | 3.3 | 32.5 | 4.0 | 2.5 | 34.2 | 63.3 | 91.1 |
| 2.5 | 3.8 | 2.4 | 34.2 | 3.5 | 2.0 | 35.9 | 71.5 | 95.1 |
| 3.0 | 3.3 | 1.9 | 35.7 | 3.2 | 1.7 | 37.2 | 77.8 | 97.3 |
| 3.5 | 3.1 | 1.7 | 36.9 | 3.1 | 1.5 | 38.3 | 82.7 | 98.5 |
| 4.0 | 3.0 | 1.5 | 37.9 | 3.0 | 1.5 | 39.2 | 86.6 | 99.2 |
| 4.5 | 2.9 | 1.4 | 38.7 | 3.0 | 1.4 | 39.8 | 89.5 | 99.6 |
| 5.0 | 2.8 | 1.4 | 39.4 | 3.0 | 1.4 | 40.4 | 91.9 | 99.8 |
| 6.0 | 2.8 | 1.3 | 40.2 | 3.0 | 1.4 | 41.0 | 95.1 | 99.9 |
| 7.0 | 2.8 | 1.3 | 40.7 | 3.0 | 1.4 | 41.4 | 97.0 | 100.0 |
| 8.0 | 2.8 | 1.3 | 41.0 | 3.0 | 1.4 | 41.7 | 98.2 | 100.0 |

APPENDIX 5 continued

2 Spherical leaf angle distribution; wet soil

| LAI | reflectance factor (%) | | | | | | soil cover (%) | |
|---|---|---|---|---|---|---|---|---|
| | sunlight | | | skylight | | | old def. | new def. |
| | G | R | IR | G | R | IR | | |
| 0.0 | 10.0 | 11.0 | 12.1 | 10.0 | 11.0 | 12.1 | 0.0 | 0.0 |
| 0.1 | 9.2 | 9.9 | 13.1 | 9.1 | 9.7 | 13.4 | 4.9 | 11.4 |
| 0.2 | 8.6 | 9.0 | 14.0 | 8.3 | 8.6 | 14.7 | 9.5 | 21.5 |
| 0.3 | 7.9 | 8.2 | 15.0 | 7.6 | 7.6 | 15.9 | 14.0 | 30.4 |
| 0.4 | 7.4 | 7.4 | 15.9 | 7.0 | 6.8 | 17.1 | 18.2 | 38.3 |
| 0.5 | 6.9 | 6.7 | 16.8 | 6.5 | 6.1 | 18.2 | 22.2 | 45.4 |
| 0.6 | 6.5 | 6.1 | 17.7 | 6.0 | 5.4 | 19.3 | 26.0 | 51.6 |

APPENDIX 5 continued

2 Spherical leaf angle distribution; wet soil

| LAI | reflectance factor (%) | | | | | | soil cover (%) | |
|---|---|---|---|---|---|---|---|---|
| | sunlight | | | skylight | | | old def. | new def. |
| | G | R | IR | G | R | IR | | |
| 0.7 | 6.1 | 5.6 | 18.6 | 5.6 | 4.9 | 20.3 | 29.6 | 57.1 |
| 0.8 | 5.7 | 5.1 | 19.5 | 5.3 | 4.4 | 21.3 | 33.1 | 62.0 |
| 0.9 | 5.4 | 4.7 | 20.3 | 5.0 | 4.0 | 22.3 | 36.3 | 66.3 |
| 1.0 | 5.1 | 4.3 | 21.1 | 4.7 | 3.7 | 23.2 | 39.5 | 70.1 |
| 1.2 | 4.6 | 3.7 | 22.7 | 4.3 | 3.1 | 25.0 | 45.2 | 76.6 |
| 1.4 | 4.3 | 3.2 | 24.2 | 4.0 | 2.7 | 26.6 | 50.5 | 81.6 |
| 1.6 | 4.0 | 2.8 | 25.7 | 3.7 | 2.4 | 28.1 | 55.2 | 85.5 |
| 1.8 | 3.7 | 2.5 | 27.0 | 3.5 | 2.1 | 29.4 | 59.5 | 88.7 |
| 2.0 | 3.5 | 2.2 | 28.2 | 3.4 | 1.9 | 30.6 | 63.3 | 91.1 |
| 2.5 | 3.2 | 1.8 | 31.0 | 3.2 | 1.7 | 33.2 | 71.5 | 95.1 |
| 3.0 | 3.0 | 1.6 | 33.3 | 3.1 | 1.5 | 35.3 | 77.8 | 97.3 |
| 3.5 | 2.9 | 1.5 | 35.1 | 3.0 | 1.5 | 36.9 | 82.7 | 98.5 |
| 4.0 | 2.9 | 1.4 | 36.5 | 3.0 | 1.4 | 38.1 | 86.6 | 99.2 |
| 4.5 | 2.8 | 1.4 | 37.7 | 3.0 | 1.4 | 39.0 | 89.5 | 99.6 |
| 5.0 | 2.8 | 1.3 | 38.6 | 3.0 | 1.4 | 39.8 | 91.9 | 99.8 |
| 6.0 | 2.8 | 1.3 | 39.8 | 3.0 | 1.4 | 40.7 | 95.1 | 99.9 |
| 7.0 | 2.8 | 1.3 | 40.5 | 3.0 | 1.4 | 41.3 | 97.0 | 100.0 |
| 8.0 | 2.8 | 1.3 | 40.9 | 3.0 | 1.4 | 41.6 | 98.2 | 100.0 |

APPENDIX 5 continued

3 Spherical leaf angle distribution; black soil

| LAI | reflectance factor (%) | | | | | | soil cover (%) | |
|---|---|---|---|---|---|---|---|---|
| | sunlight | | | skylight | | | old def. | new def. |
| | G | R | IR | G | R | IR | | |
| 0.0 | 0.0 | 0.0 | 0.0 | 0.0 | 0.0 | 0.0 | 0.0 | 0.0 |
| 0.1 | 0.3 | 0.1 | 1.7 | 0.4 | 0.2 | 2.2 | 4.9 | 11.4 |
| 0.2 | 0.5 | 0.3 | 3.3 | 0.7 | 0.3 | 4.3 | 9.5 | 21.5 |
| 0.3 | 0.8 | 0.4 | 4.9 | 1.0 | 0.5 | 6.2 | 14.0 | 30.4 |
| 0.4 | 1.0 | 0.5 | 6.4 | 1.2 | 0.6 | 8.1 | 18.2 | 38.3 |
| 0.5 | 1.2 | 0.6 | 7.8 | 1.5 | 0.7 | 9.8 | 22.2 | 45.4 |
| 0.6 | 1.3 | 0.7 | 9.2 | 1.7 | 0.8 | 11.5 | 26.0 | 51.6 |
| 0.7 | 1.5 | 0.7 | 10.6 | 1.8 | 0.9 | 13.0 | 29.6 | 57.1 |
| 0.8 | 1.6 | 0.8 | 11.9 | 2.0 | 1.0 | 14.5 | 33.1 | 62.0 |
| 0.9 | 1.8 | 0.9 | 13.2 | 2.1 | 1.0 | 15.9 | 36.3 | 66.3 |
| 1.0 | 1.9 | 0.9 | 14.4 | 2.2 | 1.1 | 17.3 | 39.5 | 70.1 |
| 1.2 | 2.1 | 1.0 | 16.7 | 2.4 | 1.2 | 19.7 | 45.2 | 76.6 |
| 1.4 | 2.2 | 1.1 | 18.9 | 2.5 | 1.2 | 22.0 | 50.5 | 81.6 |
| 1.6 | 2.3 | 1.1 | 20.9 | 2.6 | 1.3 | 24.0 | 55.2 | 85.5 |

APPENDIX 5 continued

3 Spherical leaf angle distribution; black soil

| LAI | reflectance factor (%) | | | | | | soil cover (%) | |
|---|---|---|---|---|---|---|---|---|
| | sunlight | | | skylight | | | old def. | new def. |
| | G | R | IR | G | R | IR | | |
| 1.8 | 2.4 | 1.2 | 22.7 | 2.7 | 1.3 | 25.8 | 59.5 | 88.7 |
| 2.0 | 2.5 | 1.2 | 24.4 | 2.8 | 1.3 | 27.5 | 63.3 | 91.1 |
| 2.5 | 2.6 | 1.2 | 28.1 | 2.9 | 1.4 | 30.9 | 71.5 | 95.1 |
| 3.0 | 2.7 | 1.3 | 31.1 | 2.9 | 1.4 | 33.5 | 77.8 | 97.3 |
| 3.5 | 2.7 | 1.3 | 33.4 | 2.9 | 1.4 | 35.6 | 82.7 | 98.5 |
| 4.0 | 2.8 | 1.3 | 35.3 | 2.9 | 1.4 | 37.1 | 86.6 | 99.2 |
| 4.5 | 2.8 | 1.3 | 36.7 | 2.9 | 1.4 | 38.3 | 89.5 | 99.6 |
| 5.0 | 2.8 | 1.3 | 37.9 | 3.0 | 1.4 | 39.2 | 91.9 | 99.8 |
| 6.0 | 2.8 | 1.3 | 39.4 | 3.0 | 1.4 | 40.4 | 95.1 | 99.9 |
| 7.0 | 2.8 | 1.3 | 40.3 | 3.0 | 1.4 | 41.1 | 97.0 | 100.0 |
| 8.0 | 2.8 | 1.3 | 40.8 | 3.0 | 1.4 | 41.5 | 98.2 | 100.0 |

APPENDIX 5 continued

4 Planophile leaf angle distribution; dry soil

| LAI | reflectance factor (%) | | | | | | soil cover (%) | |
|---|---|---|---|---|---|---|---|---|
| | sunlight | | | skylight | | | old def. | new def. |
| | G | R | IR | G | R | IR | | |
| 0.0 | 20.0 | 22.0 | 24.2 | 20.0 | 22.0 | 24.2 | 0.0 | 0.0 |
| 0.1 | 17.7 | 19.0 | 25.6 | 17.5 | 18.7 | 25.8 | 8.1 | 15.8 |
| 0.2 | 15.7 | 16.4 | 26.8 | 15.4 | 16.0 | 27.2 | 15.6 | 29.1 |
| 0.3 | 13.9 | 14.2 | 28.0 | 13.7 | 13.7 | 28.6 | 22.5 | 40.2 |
| 0.4 | 12.5 | 12.3 | 29.2 | 12.2 | 11.8 | 29.9 | 28.8 | 49.7 |
| 0.5 | 11.2 | 10.7 | 30.3 | 10.9 | 10.2 | 31.1 | 34.5 | 57.6 |
| 0.6 | 10.1 | 9.4 | 31.3 | 9.8 | 8.9 | 32.3 | 39.9 | 64.3 |
| 0.7 | 9.2 | 8.2 | 32.3 | 8.9 | 7.7 | 33.3 | 44.7 | 69.9 |
| 0.8 | 8.4 | 7.2 | 33.2 | 8.1 | 6.8 | 34.3 | 49.2 | 74.7 |
| 0.9 | 7.8 | 6.4 | 34.1 | 7.5 | 6.0 | 35.3 | 53.4 | 78.7 |
| 1.0 | 7.2 | 5.7 | 34.9 | 7.0 | 5.3 | 36.1 | 57.2 | 82.0 |
| 1.2 | 6.3 | 4.6 | 36.5 | 6.1 | 4.3 | 37.7 | 63.8 | 87.3 |
| 1.4 | 5.6 | 3.8 | 37.8 | 5.5 | 3.6 | 39.2 | 69.5 | 91.0 |
| 1.6 | 5.1 | 3.3 | 39.1 | 5.1 | 3.1 | 40.4 | 74.2 | 93.6 |
| 1.8 | 4.8 | 2.9 | 40.2 | 4.8 | 2.7 | 41.5 | 78.3 | 95.4 |
| 2.0 | 4.5 | 2.6 | 41.1 | 4.5 | 2.5 | 42.5 | 81.6 | 96.8 |
| 2.5 | 4.2 | 2.2 | 43.2 | 4.2 | 2.2 | 44.5 | 88.0 | 98.6 |
| 3.0 | 4.0 | 2.0 | 44.7 | 4.1 | 2.0 | 45.9 | 92.1 | 99.4 |
| 3.5 | 3.9 | 1.9 | 45.8 | 4.1 | 2.0 | 46.9 | 94.9 | 99.8 |
| 4.0 | 3.9 | 1.9 | 46.6 | 4.0 | 1.9 | 47.7 | 96.6 | 99.9 |
| 4.5 | 3.9 | 1.9 | 47.2 | 4.0 | 1.9 | 48.3 | 97.8 | 100.0 |
| 5.0 | 3.9 | 1.8 | 47.7 | 4.0 | 1.9 | 48.7 | 98.6 | 100.0 |

APPENDIX 5 continued

4 Planophile leaf angle distribution; dry soil

| LAI | reflectance factor (%) | | | | | | soil cover (%) | |
|---|---|---|---|---|---|---|---|---|
| | sunlight | | | skylight | | | old def. | new def. |
| | G | R | IR | G | R | IR | | |
| 6.0 | 3.9 | 1.8 | 48.2 | 4.0 | 1.9 | 49.2 | 99.4 | 100.0 |
| 7.0 | 3.9 | 1.8 | 48.5 | 4.0 | 1.9 | 49.5 | 99.7 | 100.0 |
| 8.0 | 3.9 | 1.8 | 48.7 | 4.0 | 1.9 | 49.6 | 99.9 | 100.0 |

APPENDIX 5 continued

5 Planophile leaf angle distribution; wet soil

| LAI | reflectance factor (%) | | | | | | soil cover (%) | |
|---|---|---|---|---|---|---|---|---|
| | sunlight | | | skylight | | | old def. | new def. |
| | G | R | IR | G | R | IR | | |
| 0.0 | 10.0 | 11.0 | 12.1 | 10.0 | 11.0 | 12.1 | 0.0 | 0.0 |
| 0.1 | 9.1 | 9.6 | 14.4 | 9.1 | 9.5 | 14.6 | 8.1 | 15.8 |
| 0.2 | 8.3 | 8.4 | 16.5 | 8.3 | 8.3 | 17.0 | 15.6 | 29.1 |
| 0.3 | 7.7 | 7.4 | 18.4 | 7.6 | 7.3 | 19.2 | 22.5 | 40.2 |
| 0.4 | 7.1 | 6.6 | 20.3 | 7.1 | 6.4 | 21.2 | 28.8 | 49.7 |
| 0.5 | 6.6 | 5.9 | 22.0 | 6.6 | 5.7 | 23.1 | 34.5 | 57.6 |
| 0.6 | 6.2 | 5.3 | 23.6 | 6.2 | 5.1 | 24.8 | 39.9 | 64.3 |
| 0.7 | 5.9 | 4.7 | 25.1 | 5.8 | 4.5 | 26.4 | 44.7 | 69.9 |
| 0.8 | 5.6 | 4.3 | 26.5 | 5.6 | 4.1 | 27.9 | 49.2 | 74.7 |
| 0.9 | 5.3 | 3.9 | 27.9 | 5.3 | 3.8 | 29.3 | 53.4 | 78.7 |
| 1.0 | 5.1 | 3.6 | 29.1 | 5.1 | 3.5 | 30.6 | 57.2 | 82.0 |
| 1.2 | 4.8 | 3.1 | 31.4 | 4.8 | 3.0 | 33.0 | 63.8 | 87.3 |
| 1.4 | 4.5 | 2.7 | 33.4 | 4.6 | 2.7 | 35.0 | 69.5 | 91.0 |
| 1.6 | 4.3 | 2.5 | 35.2 | 4.4 | 2.4 | 36.8 | 74.2 | 93.6 |
| 1.8 | 4.2 | 2.3 | 36.8 | 4.3 | 2.3 | 38.4 | 78.3 | 95.4 |
| 2.0 | 4.1 | 2.2 | 38.2 | 4.2 | 2.2 | 39.8 | 81.6 | 96.8 |
| 2.5 | 4.0 | 2.0 | 41.0 | 4.1 | 2.0 | 42.5 | 88.0 | 98.6 |
| 3.0 | 3.9 | 1.9 | 43.1 | 4.0 | 2.0 | 44.5 | 92.1 | 99.4 |
| 3.5 | 3.9 | 1.9 | 44.7 | 4.0 | 1.9 | 45.9 | 94.9 | 99.8 |
| 4.0 | 3.9 | 1.9 | 45.8 | 4.0 | 1.9 | 47.0 | 96.6 | 99.9 |
| 4.5 | 3.9 | 1.8 | 46.6 | 4.0 | 1.9 | 47.7 | 97.8 | 100.0 |
| 5.0 | 3.9 | 1.8 | 47.2 | 4.0 | 1.9 | 48.3 | 98.6 | 100.0 |
| 6.0 | 3.9 | 1.8 | 48.0 | 4.0 | 1.9 | 49.0 | 99.4 | 100.0 |
| 7.0 | 3.9 | 1.8 | 48.4 | 4.0 | 1.9 | 49.4 | 99.7 | 100.0 |
| 8.0 | 3.9 | 1.8 | 48.6 | 4.0 | 1.9 | 49.6 | 99.9 | 100.0 |

Appendix 5 continued

6 Planophile leaf angle distribution; black soil

| LAI | reflectance factor (%) | | | | | | soil cover (%) | |
|---|---|---|---|---|---|---|---|---|
| | sunlight | | | skylight | | | old def. | new def. |
| | G | R | IR | G | R | IR | | |
| 0.0 | 0.0 | 0.0 | 0.0 | 0.0 | 0.0 | 0.0 | 0.0 | 0.0 |
| 0.1 | 0.6 | 0.3 | 3.3 | 0.6 | 0.3 | 3.6 | 8.1 | 15.8 |
| 0.2 | 1.0 | 0.5 | 6.3 | 1.2 | 0.6 | 7.0 | 15.6 | 29.1 |
| 0.3 | 1.5 | 0.7 | 9.1 | 1.6 | 0.8 | 10.0 | 22.5 | 40.2 |
| 0.4 | 1.8 | 0.9 | 11.6 | 2.0 | 1.0 | 12.8 | 28.8 | 49.7 |
| 0.5 | 2.1 | 1.0 | 14.0 | 2.3 | 1.1 | 15.4 | 34.5 | 57.6 |
| 0.6 | 2.4 | 1.2 | 16.3 | 2.6 | 1.3 | 17.7 | 39.9 | 64.3 |
| 0.7 | 2.6 | 1.3 | 18.3 | 2.8 | 1.4 | 19.9 | 44.7 | 69.9 |
| 0.8 | 2.8 | 1.3 | 20.3 | 3.0 | 1.5 | 21.9 | 49.2 | 74.7 |
| 0.9 | 2.9 | 1.4 | 22.1 | 3.1 | 1.5 | 23.8 | 53.4 | 78.7 |
| 1.0 | 3.1 | 1.5 | 23.7 | 3.3 | 1.6 | 25.5 | 57.2 | 82.0 |
| 1.2 | 3.3 | 1.6 | 26.8 | 3.5 | 1.7 | 28.6 | 63.8 | 87.3 |
| 1.4 | 3.4 | 1.7 | 29.4 | 3.6 | 1.8 | 31.2 | 69.5 | 91.0 |
| 1.6 | 3.6 | 1.7 | 31.7 | 3.7 | 1.8 | 33.6 | 74.2 | 93.6 |
| 1.8 | 3.6 | 1.7 | 33.8 | 3.8 | 1.9 | 35.6 | 78.3 | 95.4 |
| 2.0 | 3.7 | 1.8 | 35.6 | 3.9 | 1.9 | 37.3 | 81.6 | 96.8 |
| 2.5 | 3.8 | 1.8 | 39.2 | 4.0 | 1.9 | 40.8 | 88.0 | 98.6 |
| 3.0 | 3.8 | 1.8 | 41.8 | 4.0 | 1.9 | 43.3 | 92.1 | 99.4 |
| 3.5 | 3.8 | 1.8 | 43.7 | 4.0 | 1.9 | 45.0 | 94.9 | 99.8 |
| 4.0 | 3.9 | 1.8 | 45.1 | 4.0 | 1.9 | 46.3 | 96.6 | 99.9 |
| 4.5 | 3.9 | 1.8 | 46.1 | 4.0 | 1.9 | 47.3 | 97.8 | 100.0 |
| 5.0 | 3.9 | 1.8 | 46.9 | 4.0 | 1.9 | 48.0 | 98.6 | 100.0 |
| 6.0 | 3.9 | 1.8 | 47.8 | 4.0 | 1.9 | 48.8 | 99.4 | 100.0 |
| 7.0 | 3.9 | 1.8 | 48.3 | 4.0 | 1.9 | 49.3 | 99.7 | 100.0 |
| 8.0 | 3.9 | 1.8 | 48.6 | 4.0 | 1.9 | 49.5 | 99.9 | 100.0 |

Appendix 5 continued

7 Erectophile leaf angle distribution; dry soil

| LAI | reflectance factor (%) | | | | | | soil cover (%) | |
|---|---|---|---|---|---|---|---|---|
| | sunlight | | | skylight | | | old def. | new def. |
| | G | R | IR | G | R | IR | | |
| 0.0 | 20.0 | 22.0 | 24.2 | 20.0 | 22.0 | 24.2 | 0.0 | 0.0 |
| 0.1 | 18.3 | 19.9 | 24.4 | 17.9 | 19.4 | 24.5 | 4.2 | 10.6 |
| 0.2 | 16.8 | 18.0 | 24.6 | 16.0 | 17.0 | 24.9 | 8.3 | 20.0 |
| 0.3 | 15.4 | 16.3 | 24.8 | 14.4 | 15.0 | 25.2 | 12.1 | 28.4 |
| 0.4 | 14.1 | 14.8 | 25.1 | 12.9 | 13.3 | 25.6 | 15.9 | 36.0 |
| 0.5 | 13.0 | 13.4 | 25.3 | 11.7 | 11.7 | 26.0 | 19.4 | 42.7 |
| 0.6 | 11.9 | 12.1 | 25.6 | 10.6 | 10.4 | 26.4 | 22.8 | 48.8 |

APPENDIX 5 continued

7 Erectophile leaf angle distribution; dry soil

| LAI | reflectance factor (%) | | | | | | soil cover (%) | |
|---|---|---|---|---|---|---|---|---|
| | sunlight | | | skylight | | | old def. | new def. |
| | G | R | IR | G | R | IR | | |
| 0.7 | 11.0 | 11.0 | 25.9 | 9.6 | 9.2 | 26.9 | 26.1 | 54.2 |
| 0.8 | 10.2 | 10.0 | 26.2 | 8.7 | 8.2 | 27.3 | 29.2 | 59.0 |
| 0.9 | 9.4 | 9.1 | 26.5 | 8.0 | 7.3 | 27.7 | 32.2 | 63.3 |
| 1.0 | 8.7 | 8.3 | 26.9 | 7.3 | 6.5 | 28.1 | 35.1 | 67.2 |
| 1.2 | 7.5 | 6.9 | 27.5 | 6.2 | 5.2 | 28.9 | 40.4 | 73.8 |
| 1.4 | 6.6 | 5.8 | 28.2 | 5.4 | 4.3 | 29.7 | 45.4 | 79.0 |
| 1.6 | 5.8 | 4.9 | 28.9 | 4.8 | 3.6 | 30.5 | 49.9 | 83.2 |
| 1.8 | 5.1 | 4.1 | 29.5 | 4.3 | 3.0 | 31.2 | 54.0 | 86.6 |
| 2.0 | 4.6 | 3.6 | 30.2 | 3.9 | 2.6 | 31.9 | 57.8 | 89.2 |
| 2.5 | 3.7 | 2.6 | 31.7 | 3.3 | 1.9 | 33.4 | 66.0 | 93.8 |
| 3.0 | 3.2 | 2.0 | 33.1 | 3.0 | 1.6 | 34.7 | 72.6 | 96.5 |
| 3.5 | 2.9 | 1.6 | 34.3 | 2.8 | 1.4 | 35.8 | 77.9 | 98.0 |
| 4.0 | 2.7 | 1.4 | 35.2 | 2.8 | 1.3 | 36.6 | 82.2 | 98.8 |
| 4.5 | 2.6 | 1.3 | 36.0 | 2.7 | 1.3 | 37.2 | 85.7 | 99.3 |
| 5.0 | 2.5 | 1.3 | 36.7 | 2.7 | 1.3 | 37.8 | 88.5 | 99.6 |
| 6.0 | 2.5 | 1.2 | 37.6 | 2.7 | 1.3 | 38.5 | 92.5 | 99.9 |
| 7.0 | 2.5 | 1.2 | 38.2 | 2.7 | 1.3 | 38.9 | 95.1 | 100.0 |
| 8.0 | 2.5 | 1.2 | 38.5 | 2.7 | 1.3 | 39.1 | 96.8 | 100.0 |

APPENDIX 5 continued

8 Erectophile leaf angle distribution; wet soil

| LAI | reflectance factor (%) | | | | | | soil cover (%) | |
|---|---|---|---|---|---|---|---|---|
| | sunlight | | | skylight | | | old def. | new def. |
| | G | R | IR | G | R | IR | | |
| 0.0 | 10.0 | 11.0 | 12.1 | 10.0 | 11.0 | 12.1 | 0.0 | 0.0 |
| 0.1 | 9.3 | 10.0 | 12.8 | 9.1 | 9.8 | 13.1 | 4.2 | 10.6 |
| 0.2 | 8.6 | 9.1 | 13.5 | 8.3 | 8.7 | 14.2 | 8.3 | 20.0 |
| 0.3 | 8.0 | 8.3 | 14.3 | 7.6 | 7.7 | 15.2 | 12.1 | 28.4 |
| 0.4 | 7.5 | 7.6 | 15.0 | 7.0 | 6.9 | 16.2 | 15.9 | 36.0 |
| 0.5 | 7.0 | 6.9 | 15.7 | 6.5 | 6.2 | 17.1 | 19.4 | 42.7 |
| 0.6 | 6.5 | 6.3 | 16.4 | 6.0 | 5.5 | 18.0 | 22.8 | 48.8 |
| 0.7 | 6.1 | 5.8 | 17.2 | 5.6 | 5.0 | 18.9 | 26.1 | 54.2 |
| 0.8 | 5.8 | 5.3 | 17.9 | 5.2 | 4.5 | 19.8 | 29.2 | 59.0 |
| 0.9 | 5.4 | 4.9 | 18.6 | 4.9 | 4.1 | 20.7 | 32.2 | 63.3 |
| 1.0 | 5.1 | 4.5 | 19.3 | 4.6 | 3.7 | 21.5 | 35.1 | 67.2 |
| 1.2 | 4.6 | 3.9 | 20.7 | 4.2 | 3.1 | 23.0 | 40.4 | 73.8 |
| 1.4 | 4.2 | 3.3 | 22.0 | 3.8 | 2.7 | 24.5 | 45.4 | 79.0 |

Appendix 5 continued

8 Erectophile leaf angle distribution; wet soil

| LAI | reflectance factor (%) | | | | | | soil cover (%) | |
|---|---|---|---|---|---|---|---|---|
| | sunlight | | | skylight | | | old def. | new def. |
| | G | R | IR | G | R | IR | | |
| 1.6 | 3.9 | 2.9 | 23.2 | 3.5 | 2.3 | 25.8 | 49.9 | 83.2 |
| 1.8 | 3.6 | 2.6 | 24.5 | 3.3 | 2.1 | 27.0 | 54.0 | 86.6 |
| 2.0 | 3.4 | 2.3 | 25.6 | 3.2 | 1.9 | 28.2 | 57.8 | 89.2 |
| 2.5 | 3.0 | 1.8 | 28.2 | 2.9 | 1.6 | 30.6 | 66.0 | 93.8 |
| 3.0 | 2.8 | 1.5 | 30.4 | 2.8 | 1.4 | 32.6 | 72.6 | 96.5 |
| 3.5 | 2.6 | 1.4 | 32.2 | 2.7 | 1.3 | 34.1 | 77.9 | 98.0 |
| 4.0 | 2.6 | 1.3 | 33.6 | 2.7 | 1.3 | 35.4 | 82.2 | 98.8 |
| 4.5 | 2.5 | 1.2 | 34.8 | 2.7 | 1.3 | 36.3 | 85.7 | 99.3 |
| 5.0 | 2.5 | 1.2 | 35.8 | 2.7 | 1.3 | 37.1 | 88.5 | 99.6 |
| 6.0 | 2.5 | 1.2 | 37.1 | 2.7 | 1.3 | 38.1 | 92.5 | 99.9 |
| 7.0 | 2.5 | 1.2 | 37.9 | 2.7 | 1.3 | 38.7 | 95.1 | 100.0 |
| 8.0 | 2.5 | 1.2 | 38.3 | 2.7 | 1.3 | 39.0 | 96.8 | 100.0 |

Appendix 5 continued

9 Erectophile leaf angle distribution; black soil

| LAI | reflectance factor (%) | | | | | | soil cover (%) | |
|---|---|---|---|---|---|---|---|---|
| | sunlight | | | skylight | | | old def. | new def. |
| | G | R | IR | G | R | IR | | |
| 0.0 | 0.0 | 0.0 | 0.0 | 0.0 | 0.0 | 0.0 | 0.0 | 0.0 |
| 0.1 | 0.2 | 0.1 | 1.4 | 0.3 | 0.2 | 1.9 | 4.2 | 10.6 |
| 0.2 | 0.4 | 0.2 | 2.7 | 0.6 | 0.3 | 3.7 | 8.3 | 20.0 |
| 0.3 | 0.6 | 0.3 | 4.0 | 0.9 | 0.4 | 5.4 | 12.1 | 28.4 |
| 0.4 | 0.8 | 0.4 | 5.3 | 1.1 | 0.5 | 7.0 | 15.9 | 36.0 |
| 0.5 | 1.0 | 0.5 | 6.5 | 1.3 | 0.6 | 8.6 | 19.4 | 42.7 |
| 0.6 | 1.1 | 0.6 | 7.7 | 1.4 | 0.7 | 10.0 | 22.8 | 48.8 |
| 0.7 | 1.3 | 0.6 | 8.9 | 1.6 | 0.8 | 11.4 | 26.1 | 54.2 |
| 0.8 | 1.4 | 0.7 | 10.1 | 1.7 | 0.8 | 12.8 | 29.2 | 59.0 |
| 0.9 | 1.5 | 0.7 | 11.2 | 1.8 | 0.9 | 14.1 | 32.2 | 63.3 |
| 1.0 | 1.6 | 0.8 | 12.3 | 1.9 | 0.9 | 15.3 | 35.1 | 67.2 |
| 1.2 | 1.7 | 0.8 | 14.3 | 2.1 | 1.0 | 17.6 | 40.4 | 73.8 |
| 1.4 | 1.9 | 0.9 | 16.3 | 2.2 | 1.1 | 19.6 | 45.4 | 79.0 |
| 1.6 | 2.0 | 1.0 | 18.1 | 2.3 | 1.1 | 21.5 | 49.9 | 83.2 |
| 1.8 | 2.1 | 1.0 | 19.8 | 2.4 | 1.2 | 23.2 | 54.0 | 86.6 |
| 2.0 | 2.2 | 1.0 | 21.4 | 2.5 | 1.2 | 24.8 | 57.8 | 89.2 |
| 2.5 | 2.3 | 1.1 | 25.0 | 2.6 | 1.2 | 28.1 | 66.0 | 93.8 |
| 3.0 | 2.4 | 1.1 | 27.9 | 2.6 | 1.2 | 30.7 | 72.6 | 96.5 |
| 3.5 | 2.4 | 1.1 | 30.3 | 2.6 | 1.3 | 32.7 | 77.9 | 98.0 |
| 4.0 | 2.4 | 1.2 | 32.2 | 2.7 | 1.3 | 34.3 | 82.2 | 98.8 |

APPENDIX 5 continued

9 Erectophile leaf angle distribution; black soil

| LAI | reflectance factor (%) | | | | | | soil cover (%) | |
|---|---|---|---|---|---|---|---|---|
| | sunlight | | | skylight | | | old def. | new def. |
| | G | R | IR | G | R | IR | | |
| 4.5 | 2.4 | 1.2 | 33.7 | 2.7 | 1.3 | 35.5 | 85.7 | 99.3 |
| 5.0 | 2.5 | 1.2 | 34.9 | 2.7 | 1.3 | 36.5 | 88.5 | 99.6 |
| 6.0 | 2.5 | 1.2 | 36.6 | 2.7 | 1.3 | 37.8 | 92.5 | 99.9 |
| 7.0 | 2.5 | 1.2 | 37.6 | 2.7 | 1.3 | 38.5 | 95.1 | 100.0 |
| 8.0 | 2.5 | 1.2 | 38.2 | 2.7 | 1.3 | 38.9 | 96.8 | 100.0 |

APPENDIX 6 Soil cover (new definition) as a function of green and red reflectance, respectively, for a planophile and an erectophile leaf angle distribution.
xx : calculated points SAIL model
— : simplified reflectance model

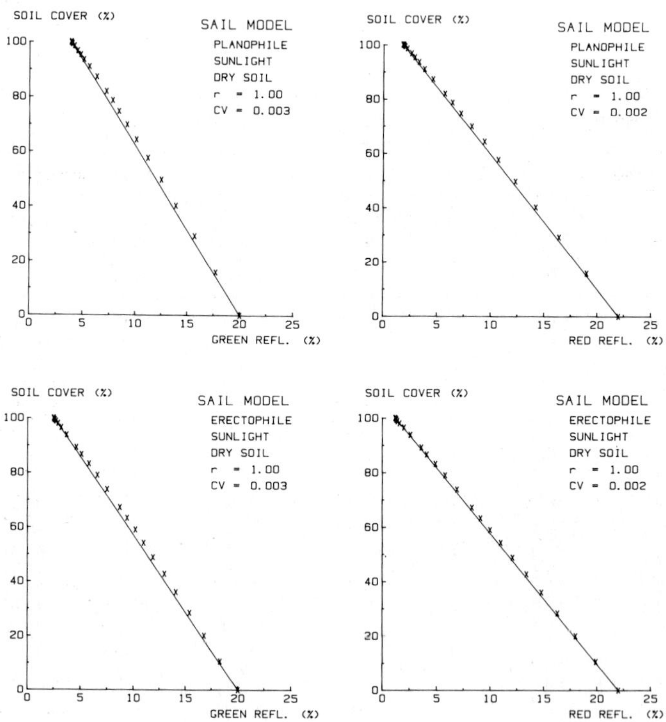

APPENDIX 7 Three methods for correcting for differences in soil moisture content in estimating soil cover for a dry and a wet soil, with a planophile and an erectophile leaf angle distribution, respectively.

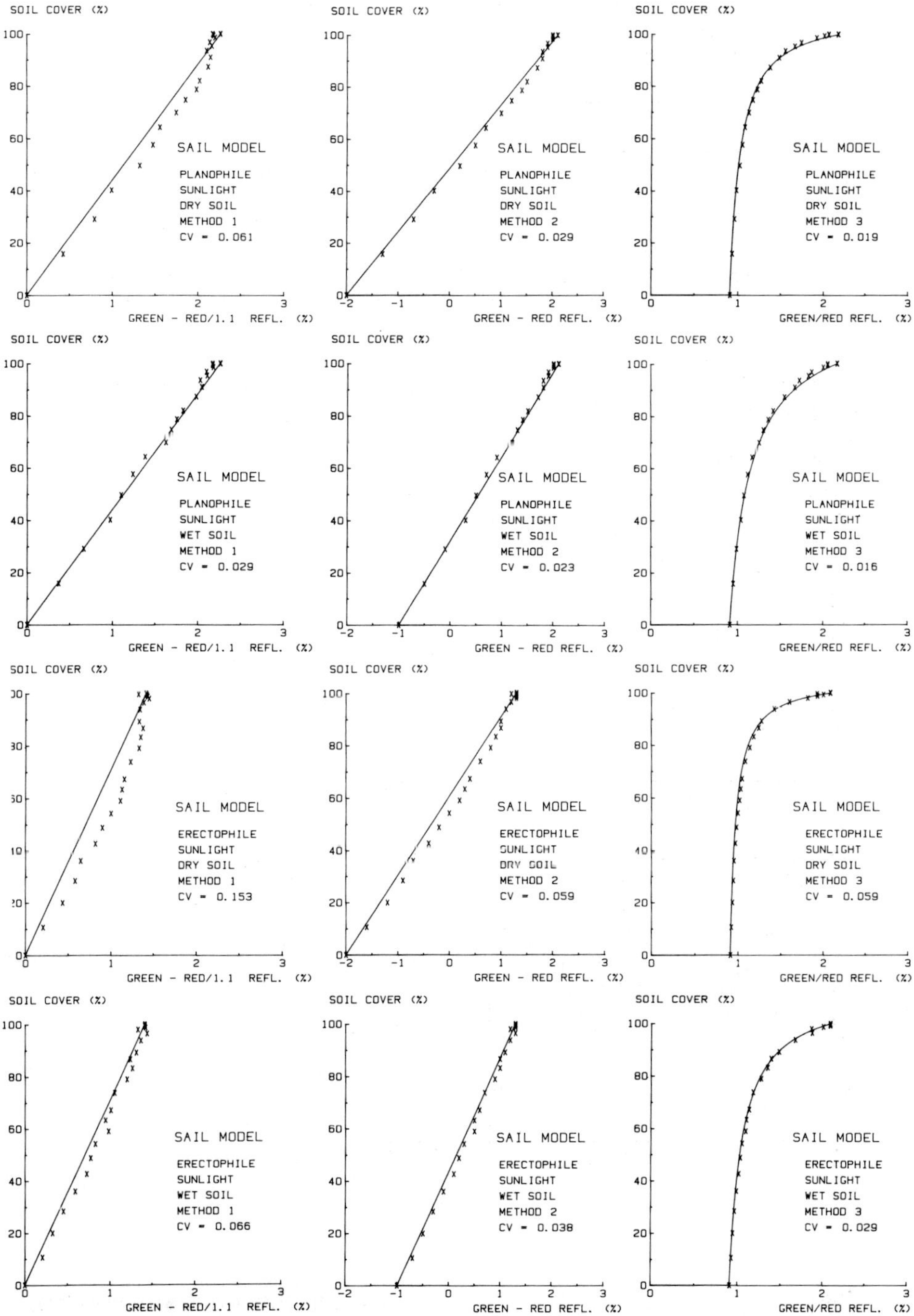

APPENDIX 8 LAI as a function of the infrared reflectance for a black soil, with a planophile and an erectophile leaf angle distribution, respectively.
xx : calculated points SAIL model
— : simplified reflectance model

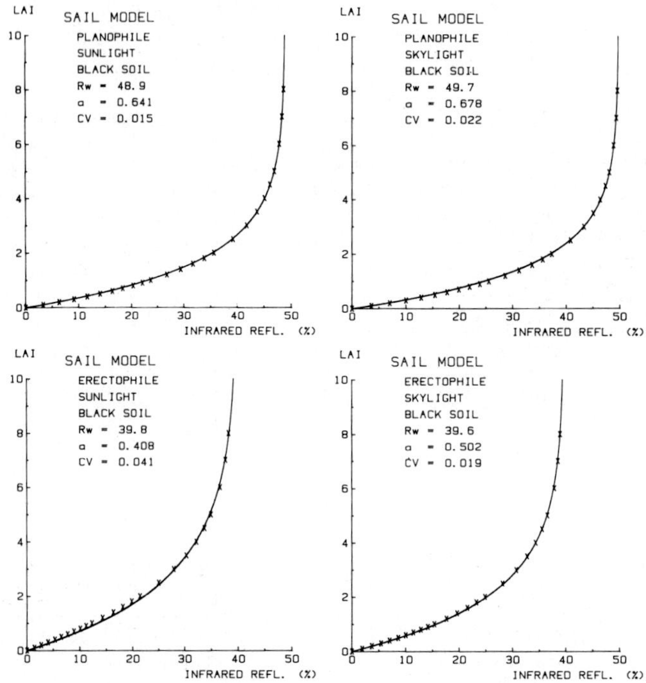

APPENDIX 9 Three methods for correcting for differences in soil moisture content in estimating LAI. Planophile leaf angle distribution.

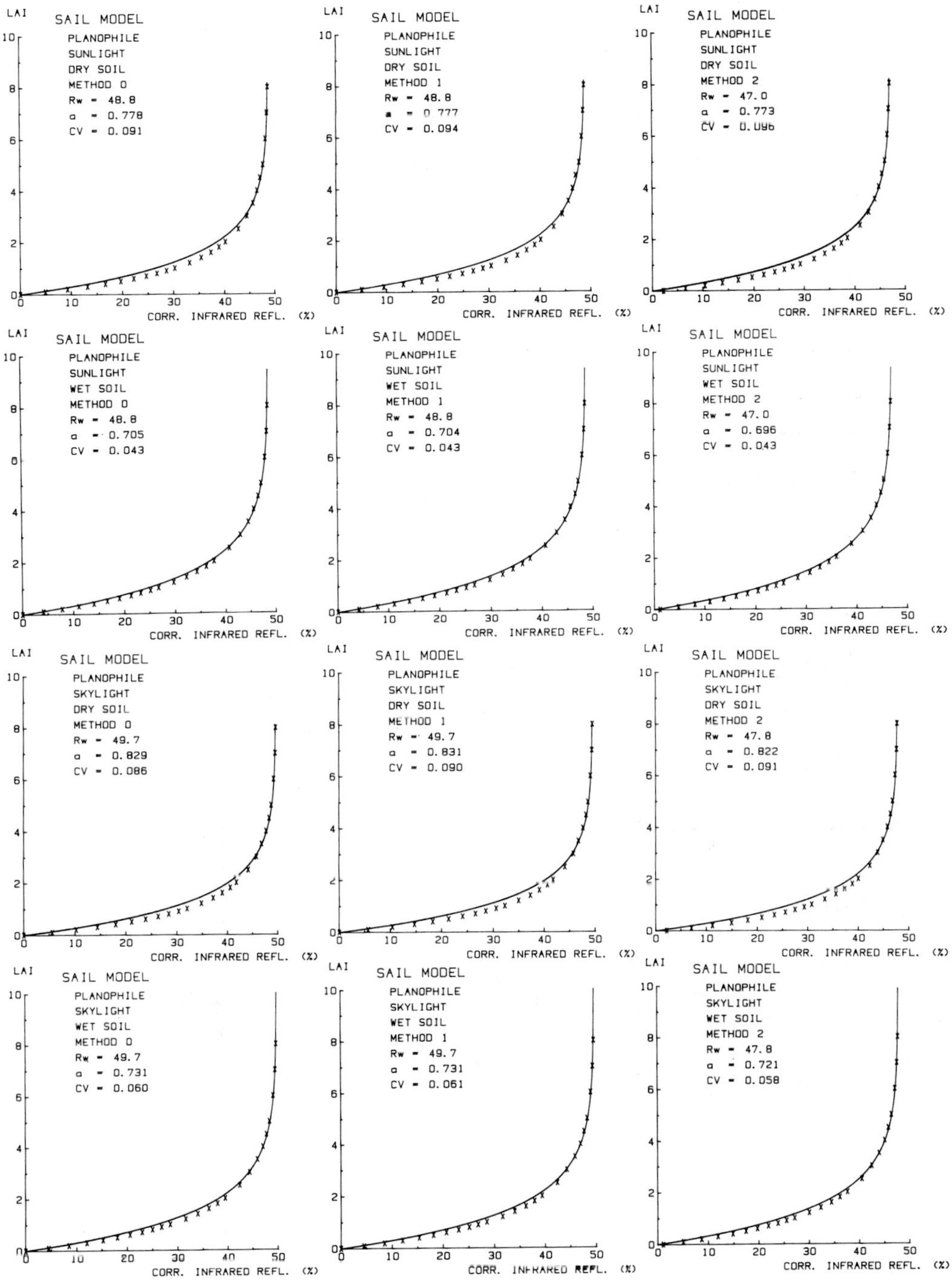

APPENDIX 10 Three methods for correcting for differences in soil moisture content in estimating LAI. Erectophile leaf angle distribution.

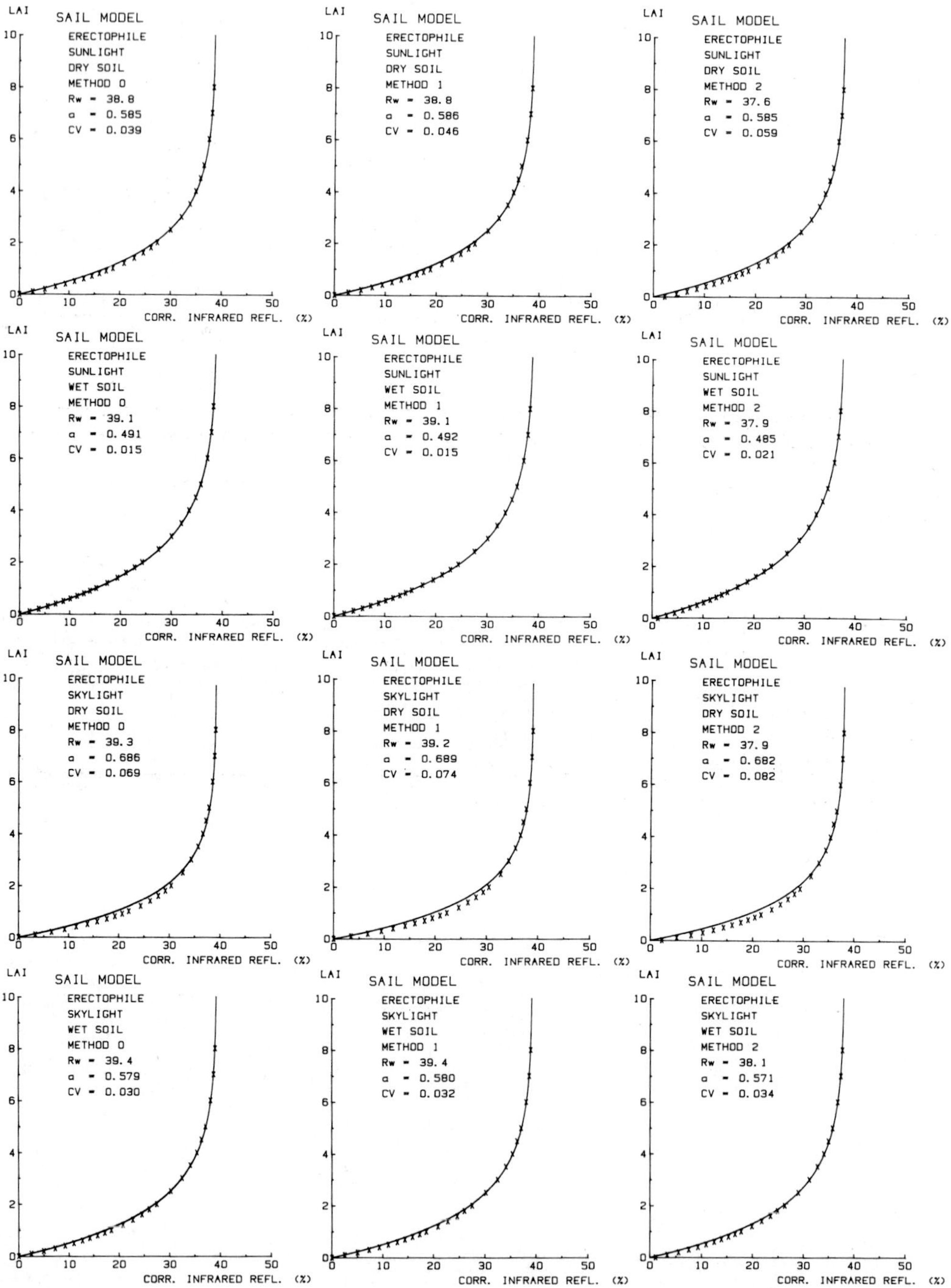

APPENDIX 11A Daily precipitation and minimum/maximum temperatures for the 1982 growing season. Days of missions are indicated by arrows.

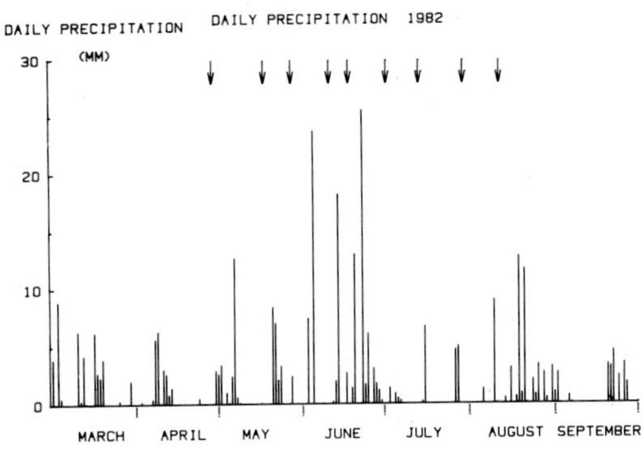


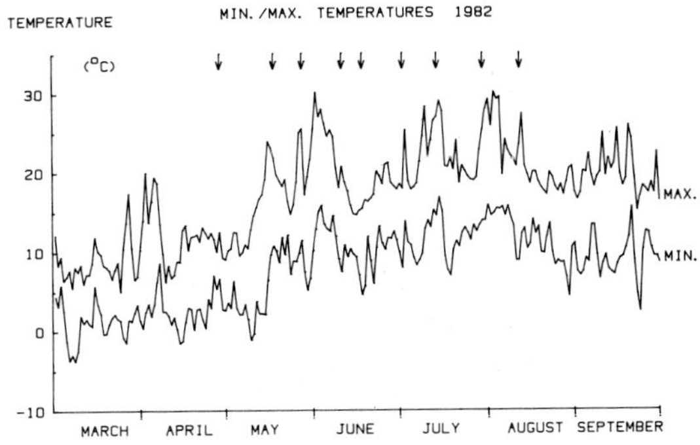


APPENDIX 11B Daily precipitation and minimum/maximum temperatures for the 1983 growing season. 11–31 July: no rainfall data, because of computer failure. Days of missions are indicated by arrows.

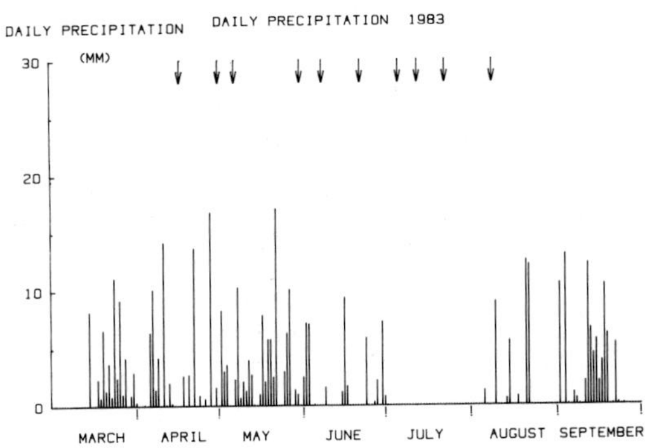


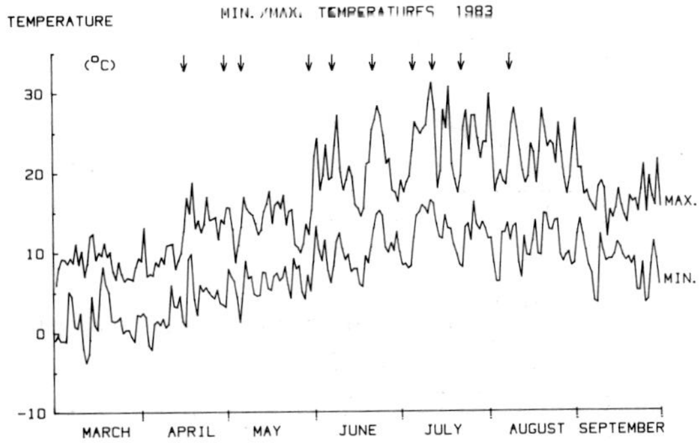

APPENDIX 12 LAI per harvesting date. Field trial 92 in 1982.

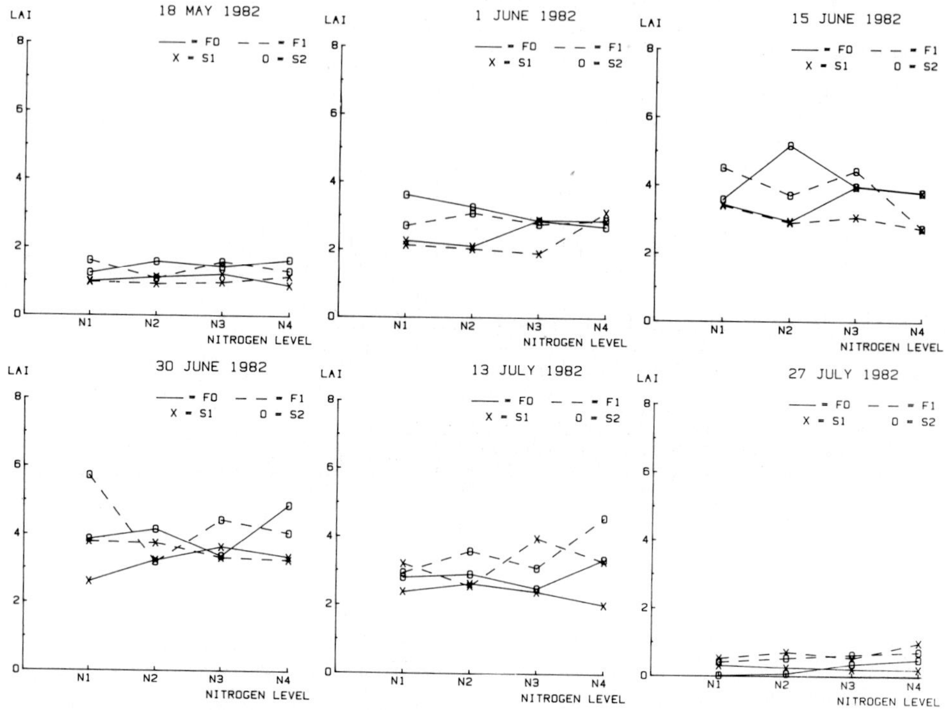


APPENDIX 13 LAI per harvesting date. Field trial 92 in 1983.

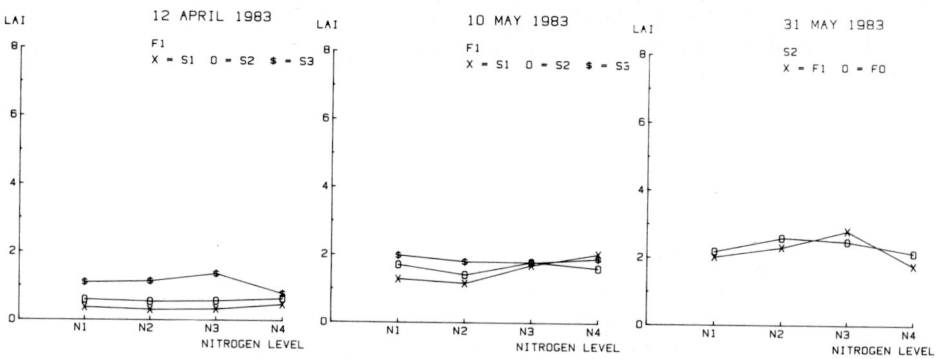

APPENDIX 13 continued.

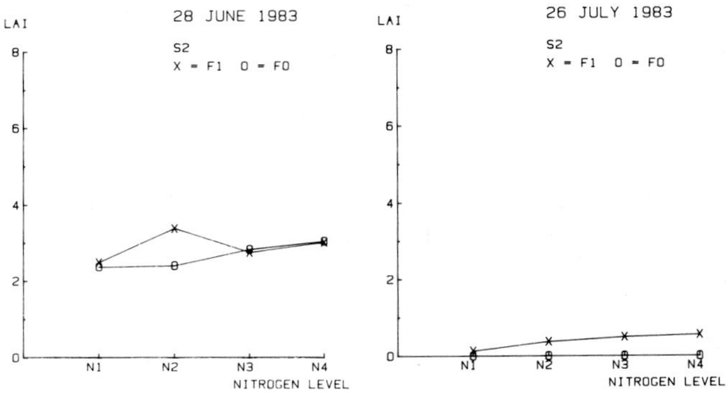


APPENDIX 14 LAI per harvesting date. Field trial 95 in 1983.

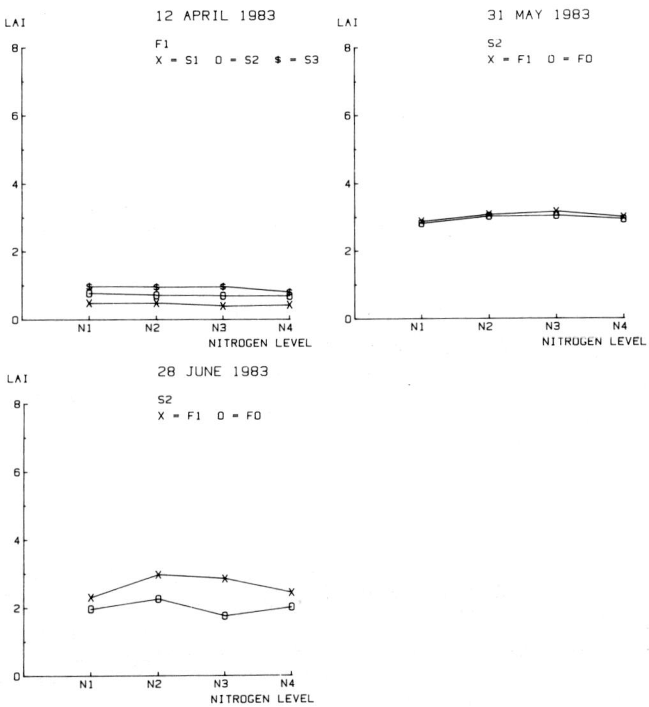

APPENDIX 15: Dry matter weight: means, CVs and critical levels in testing for treatment effects, obtained on consecutive sampling dates.
Field trial 116 in 1982 (36 plots)

| date | mean (g/m²) | CV | critical level in testing: | | |
|---|---|---|---|---|---|
| | | | inter-action | sowing dates | nitrogen nutrition |
| 19 May | 87 | 0.216 | 0.029 | 0.002 | 0.010 |
| 26 May | 173 | 0.163 | 0.165 | 0.000 | 0.020 |
| 3 June | 369 | 0.171 | 0.041 | 0.003 | 0.002 |
| 10 June | 577 | 0.114 | 0.107 | 0.001 | 0.000 |
| 17 June | 664 | 0.222 | 0.768 | 0.001 | 0.152 |
| 25 June | 925 | 0.173 | 0.492 | 0.003 | 0.007 |
| 2 July | 1215 | 0.209 | 0.765 | 0.131 | 0.142 |
| 8 July | 1510 | 0.230 | 0.013 | 0.423 | 0.409 |
| 15 July | 1561 | 0.134 | 0.492 | 0.086 | 0.001 |
| 21 July | 1527 | 0.129 | 0.508 | 0.021 | 0.001 |
| 27 July | 1650 | 0.133 | 0.132 | 0.198 | 0.005 |
| 10 August | 1564 | 0.152 | 0.723 | 0.264 | 0.040 |

Field trial 116 in 1983 (40 plots)

| date | mean (g/m²) | CV | critical level in testing: | | |
|---|---|---|---|---|---|
| | | | inter-action | sowing dates | nitrogen nutrition |
| 10 May | 45 | 0.241 | 0.562 | 0.000 | 0.549 |
| 31 May | 191 | 0.233 | 0.055 | 0.002 | 0.002 |
| 6 June | 463 | 0.157 | 0.204 | 0.001 | 0.001 |
| 21 June | 510 | 0.181 | 0.114 | 0.001 | 0.002 |
| 28 June | 847 | 0.166 | 0.267 | 0.001 | 0.001 |
| 5 July | 791 | 0.176 | 0.009 | 0.010 | 0.005 |
| 12 July | 1066 | 0.184 | 0.081 | 0.016 | 0.018 |
| 19 July | 1109 | 0.187 | 0.224 | 0.219 | 0.003 |
| 2 August | 1146 | 0.170 | 0.744 | 0.017 | 0.015 |

APPENDIX 16: Occurrence of development stages of barley.
Field trial 116 in 1982.

| Feekes scale | | date of occurrence | |
|---|---|---|---|
| | | Z1 | Z2 |
| F1 | (first shoot) | 30 April | 15 May |
| F3 | (tillering) | 15 May | 28 May |
| F5 | (end tillering) | 25 May | 5 June |
| F7 | (first node) | 5 June | 15 June |
| F10 | (last leaf) | 15 June | 25 June |
| F10.5 | (ear visible) | 25 June | 5 July |
| F11 | (ripening) | 15 July | 20 July |

APPENDIX 16 continued.

Field trial 116 in 1983.

| Feekes scale | | date of occurrence | |
|---|---|---|---|
| | | Z1 | Z2 |
| F1 | (first shoot) | 15 April | 15 May |
| F3 | (tillering) | 5 May | 30 May |
| F5 | (end tillering) | 25 May | 5 June |
| F7 | (first node) | 5 June | 15 June |
| F10 | (last leaf) | 15 June | 25 June |
| F10.5 | (ear visible) | 25 June | 5 July |
| F11 | (ripening) | 15 July | 20 July |

APPENDIX 17: Occurrence of development stages of wheat.
Field trial 92 in 1982 (spring wheat)

| Feekes scale | | date of occurrence |
|---|---|---|
| F1 | (first shoot) | 30 April |
| F3 | (tillering) | 15 May |
| F5 | (end tillering) | 25 May |
| F7 | (first node) | 5 June |
| F10 | (last leaf) | 15 June |
| F10.5 | (ear visible) | 20 June |
| F11 | (ripening) | 10 July |

Field trials 92 and 95 in 1983 (winter wheat)

| Feekes scale | | date of occurrence |
|---|---|---|
| F1 | (first shoot) | 30 November (1982) |
| F3 | (tillering) | 20 April |
| F5 | (end tillering) | 10 May |
| F7 | (first node) | 20 May |
| F10 | (last leaf) | 5 June |
| F10.5 | (ear visible) | 15 June |
| F11 | (ripening) | 5 July |

APPENDIX 18 Green and red reflectances per flying date. Field trial 116 in 1983.

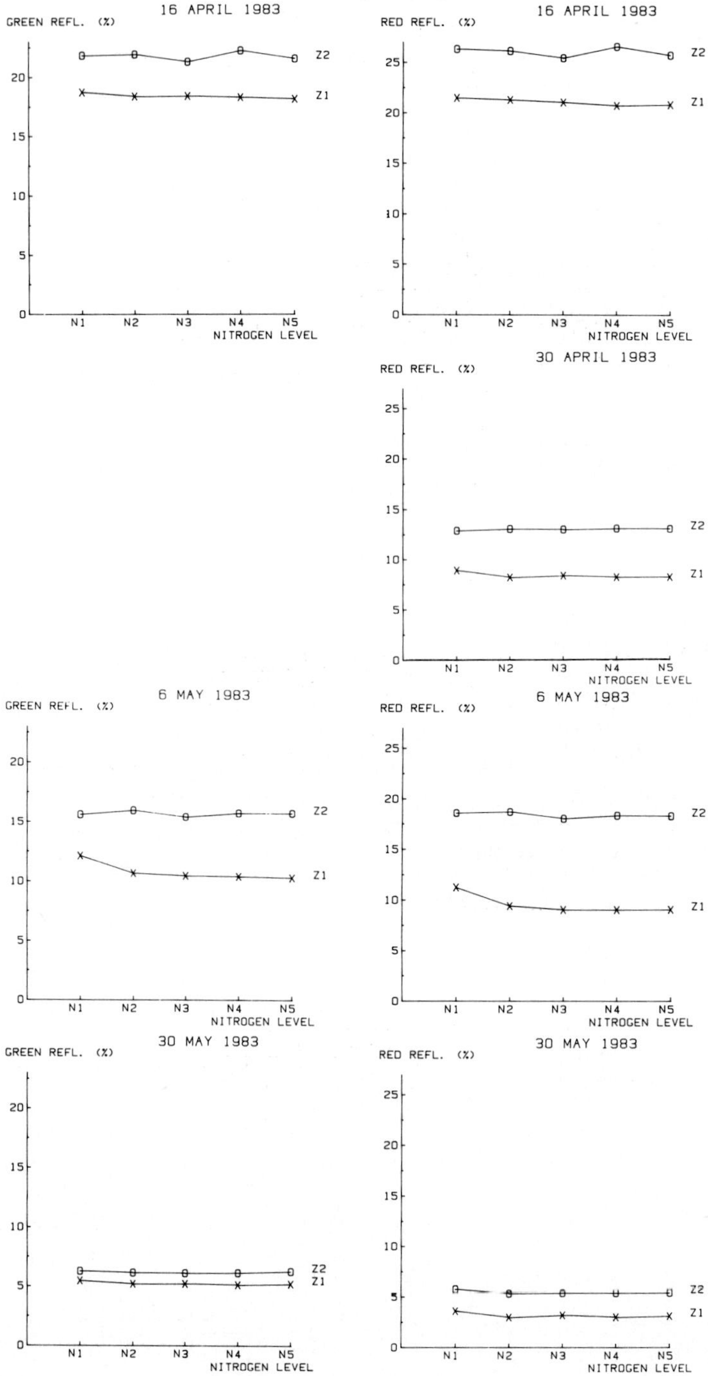

APPENDIX 18 continued.

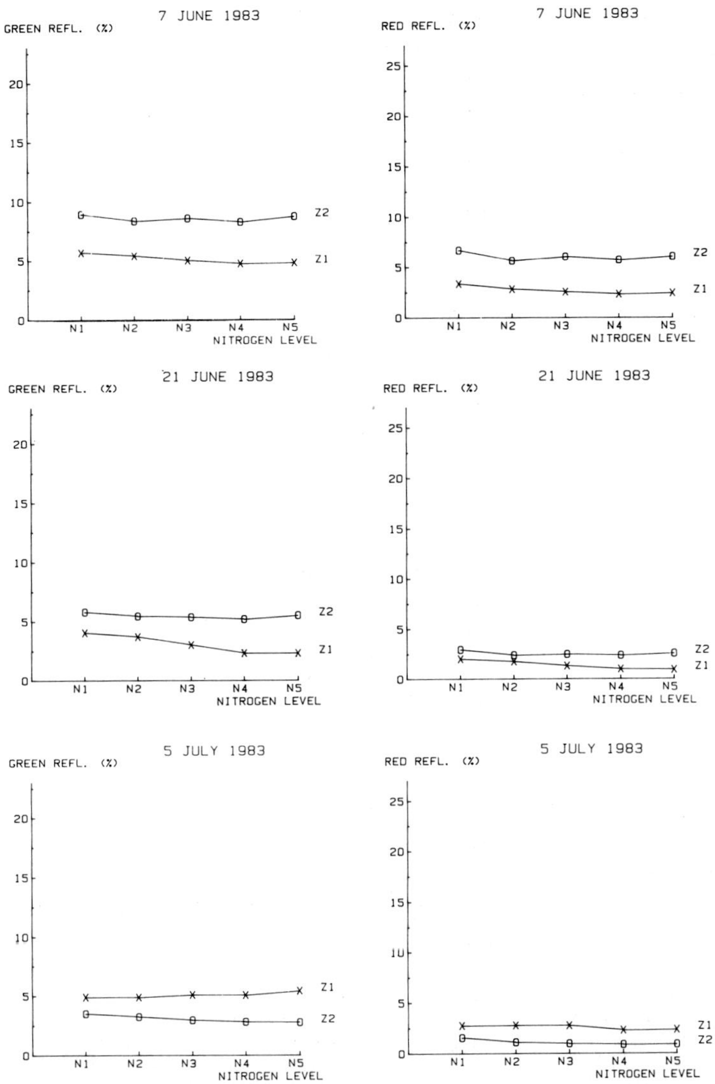

APPENDIX 18 continued.

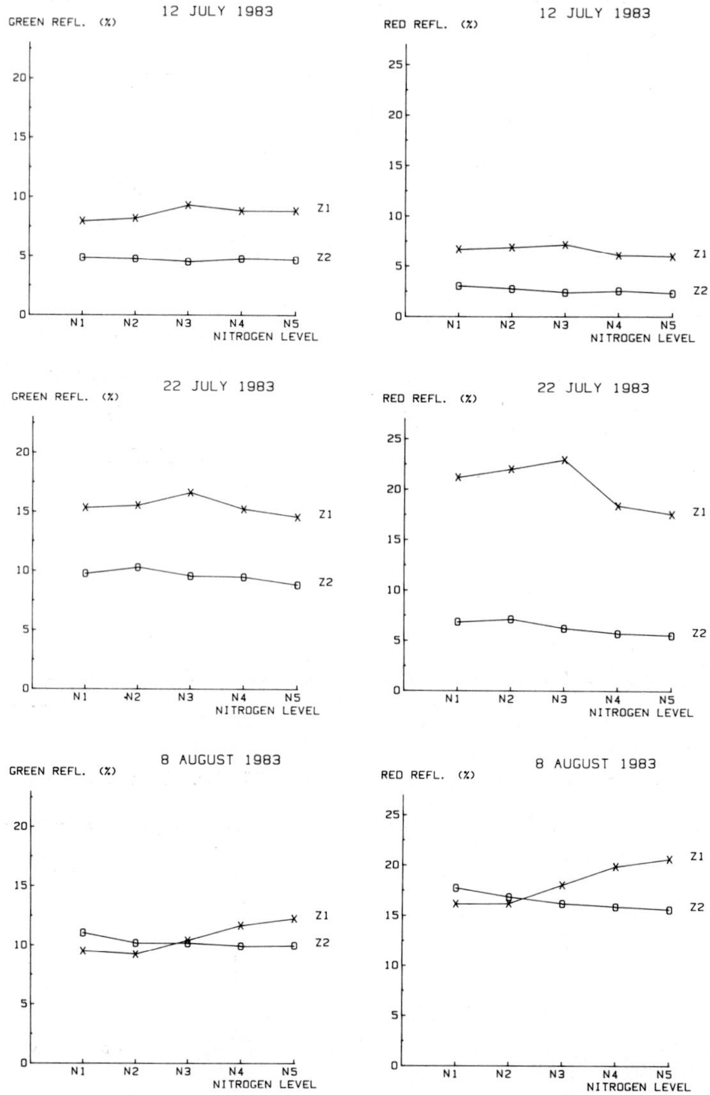

APPENDIX 19 Green and red reflectances as a function of time. Field trial 116 in 1983.

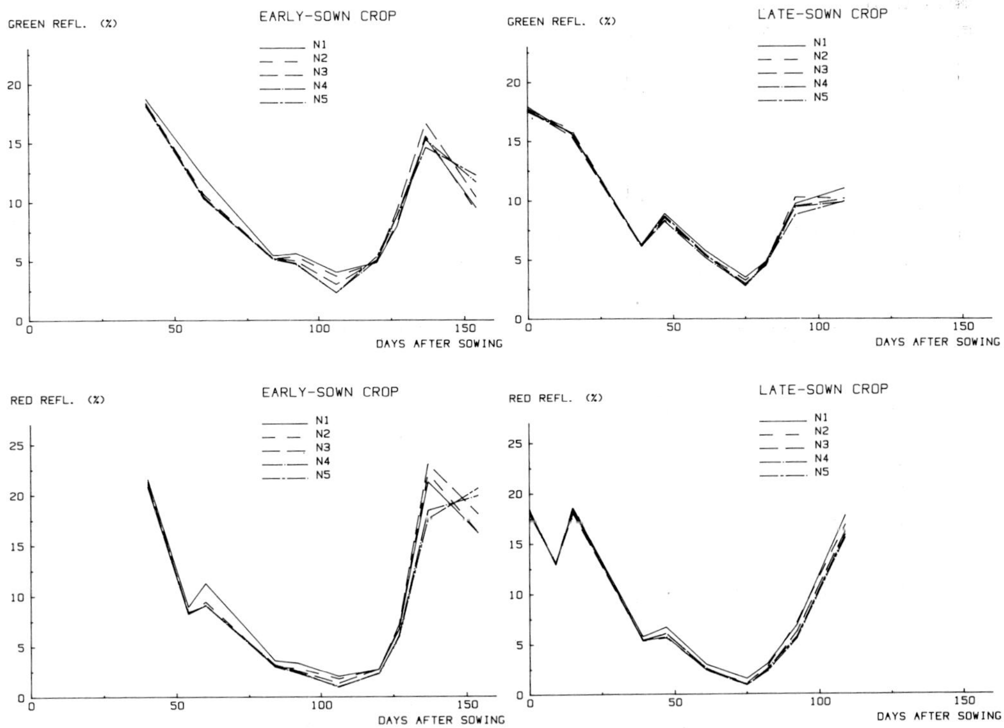


APPENDIX 20 Green and red and infrared reflectances per flying date. Field trial 92 in 1982.

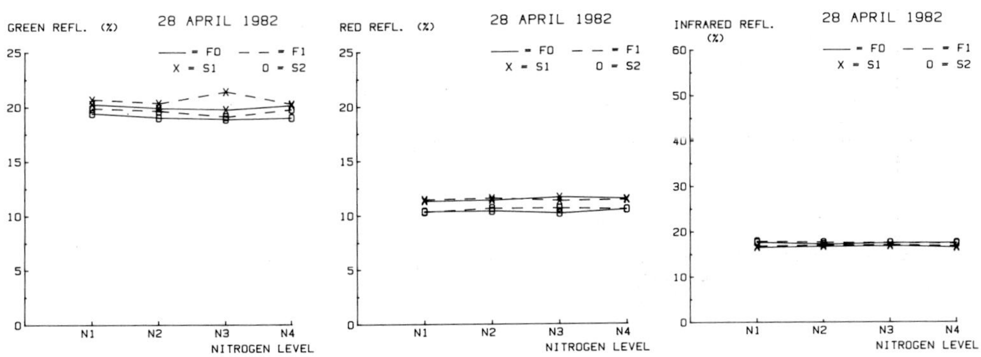

APPENDIX 20 continued.

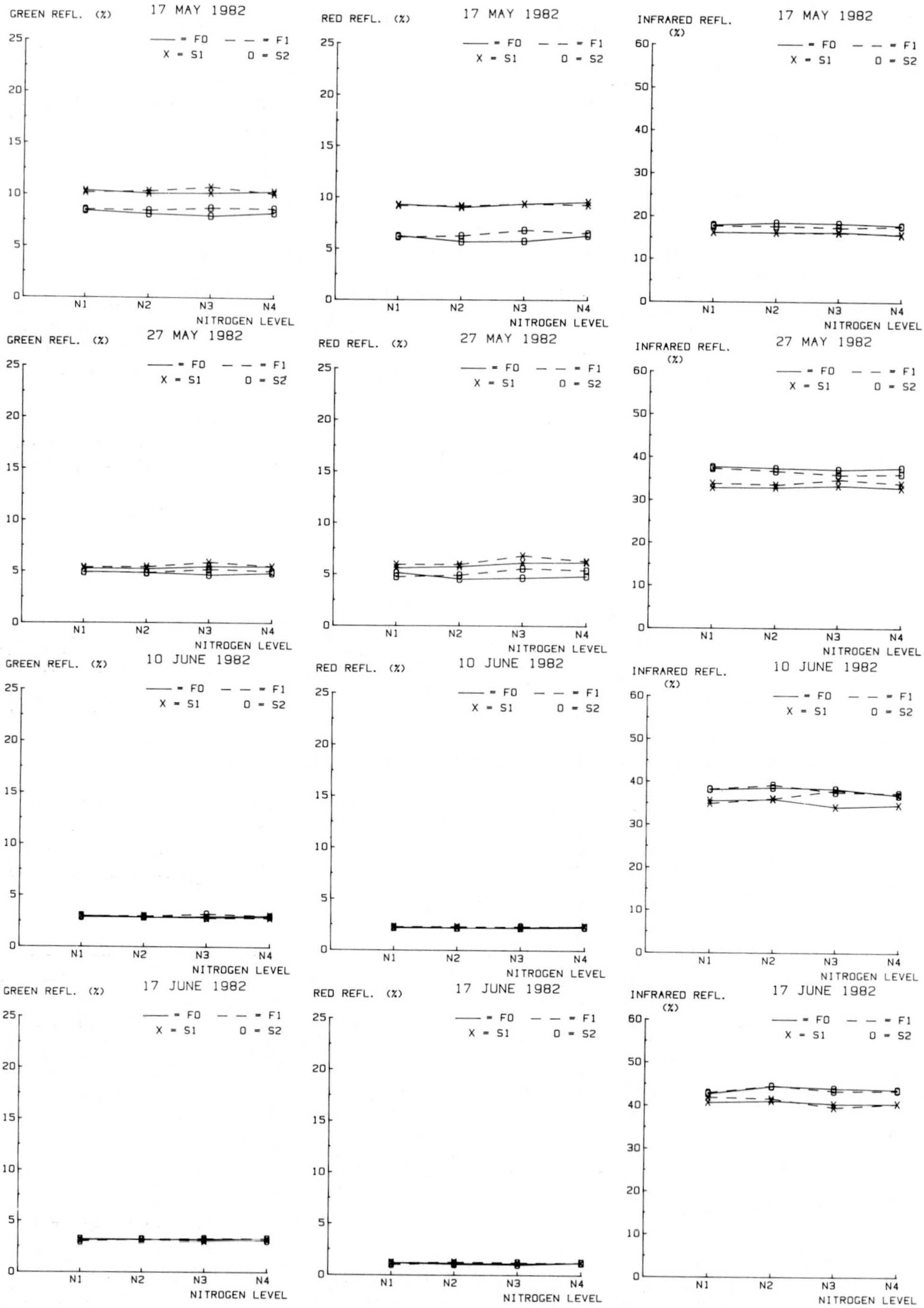

Appendix 20 continued.

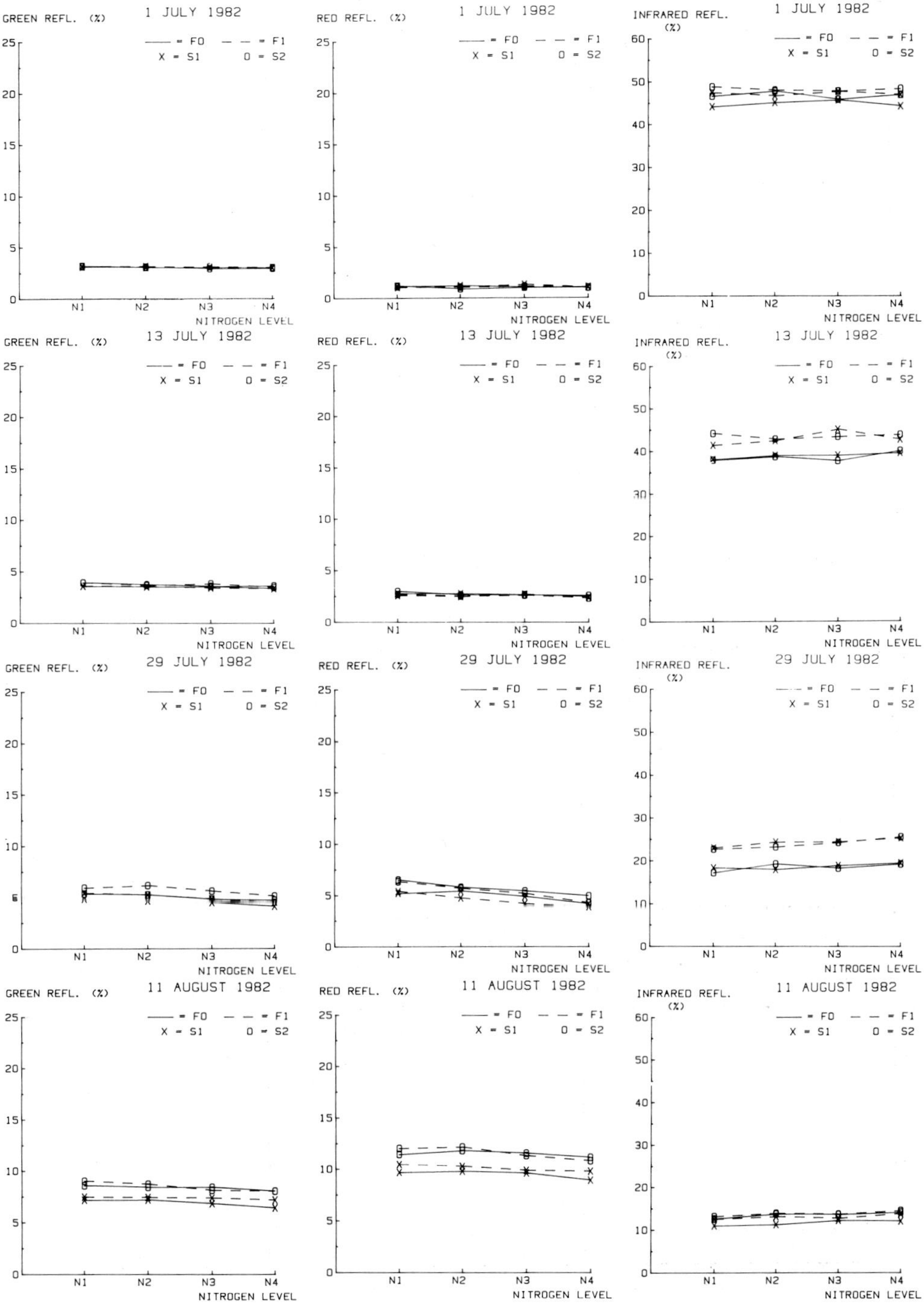

APPENDIX 21 Green and red and infrared reflectances per flying date. Field trial 92 in 1983.

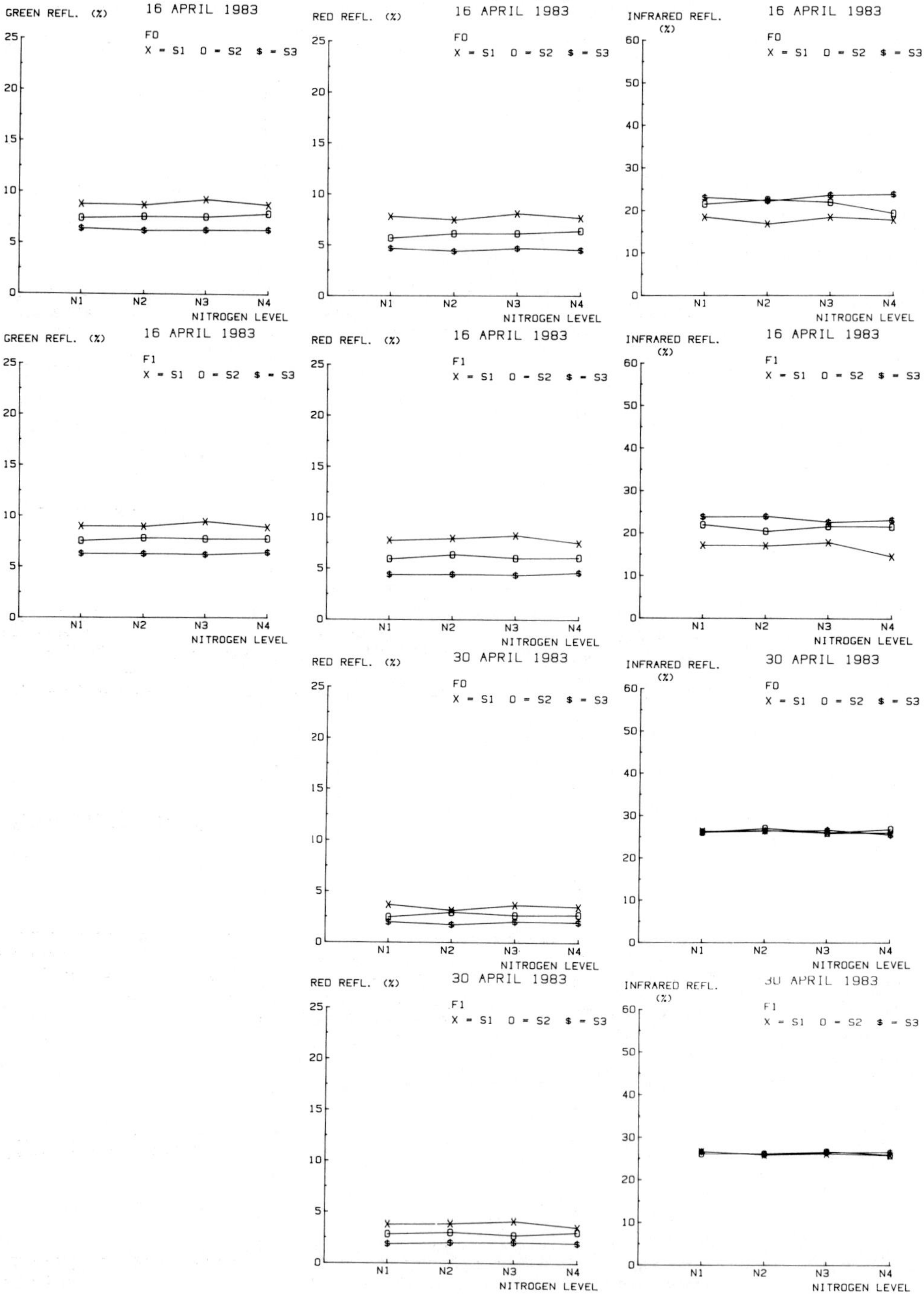

Appendix 21 continued.

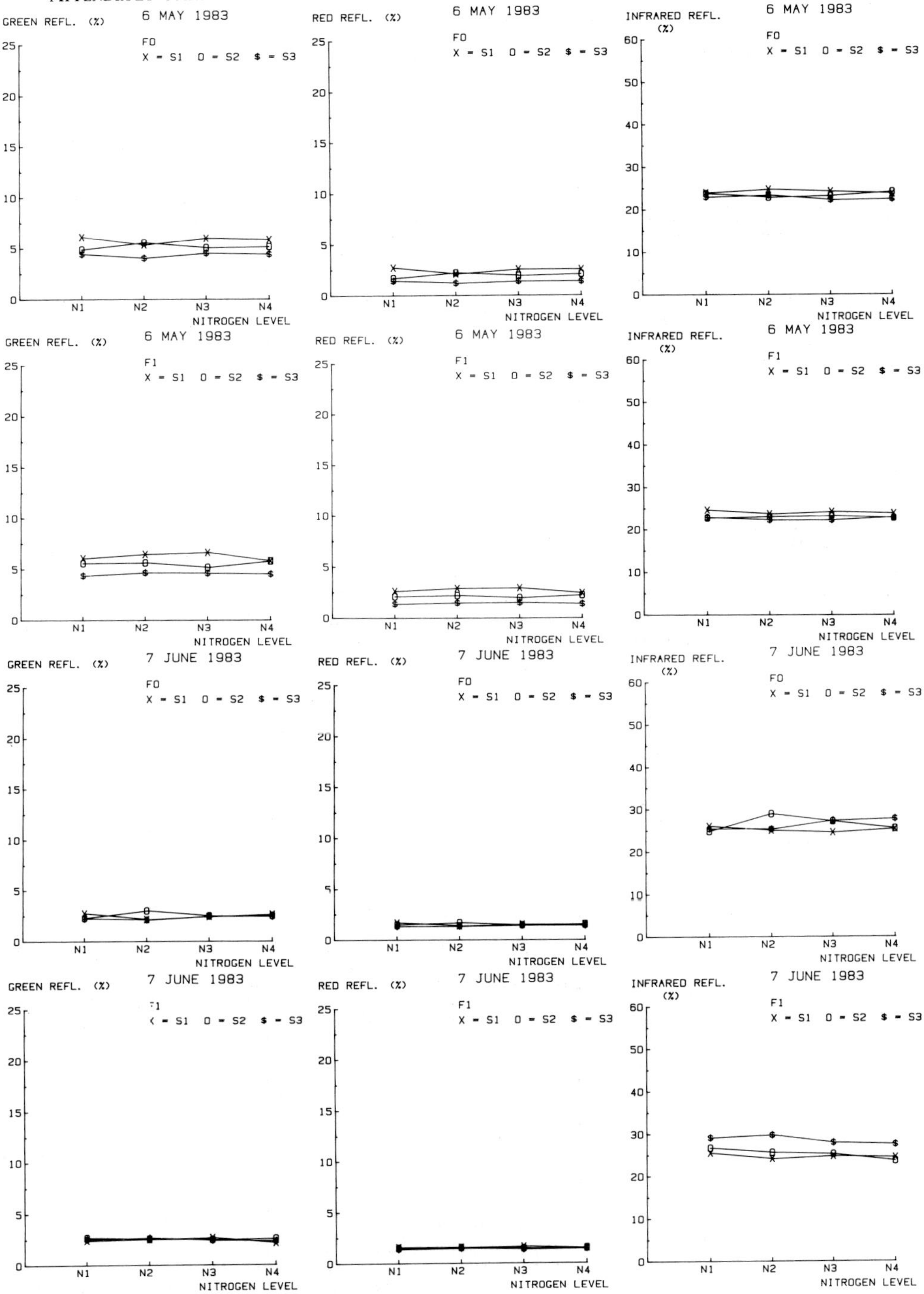

APPENDIX 21 continued.

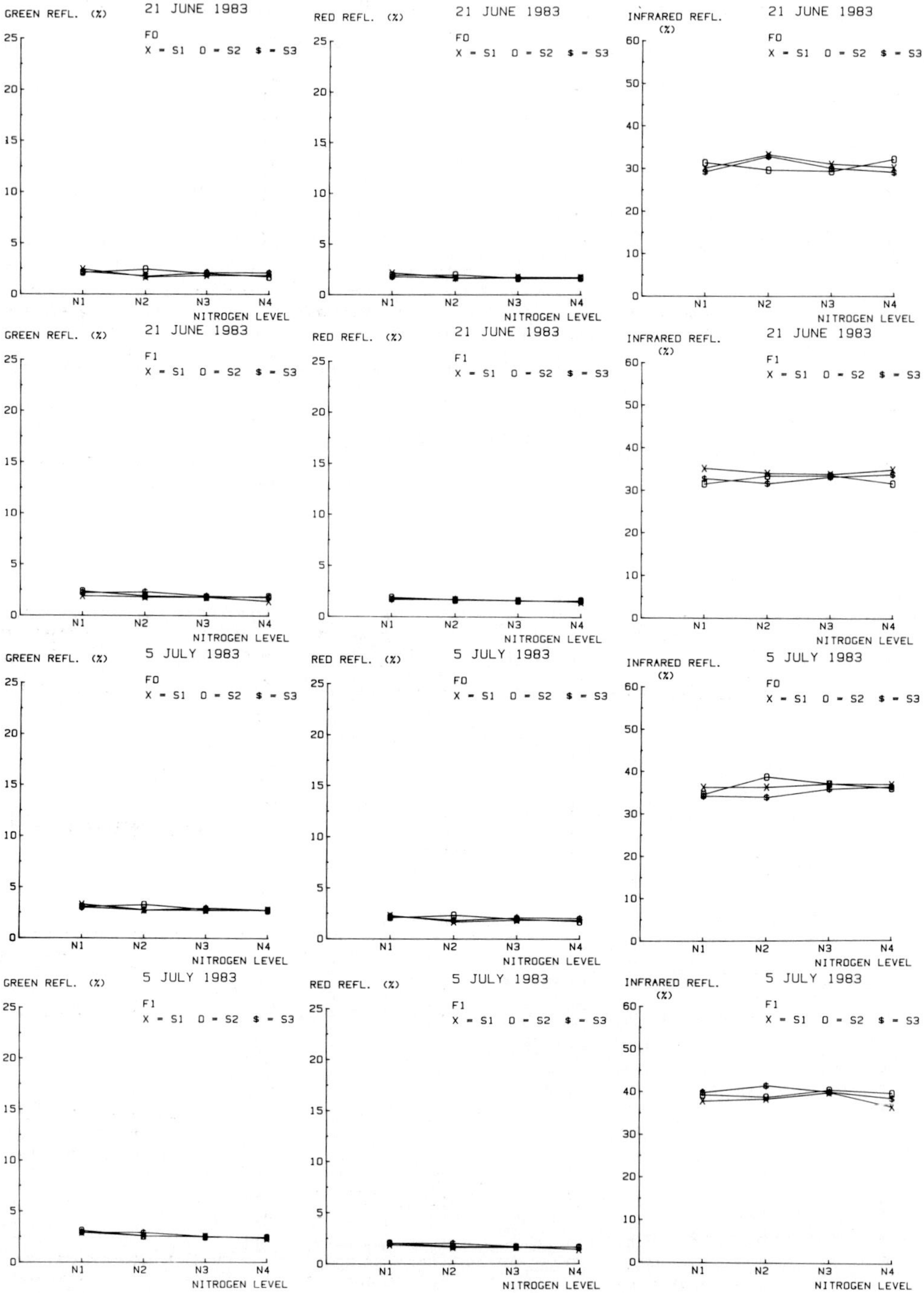

APPENDIX 21 continued.

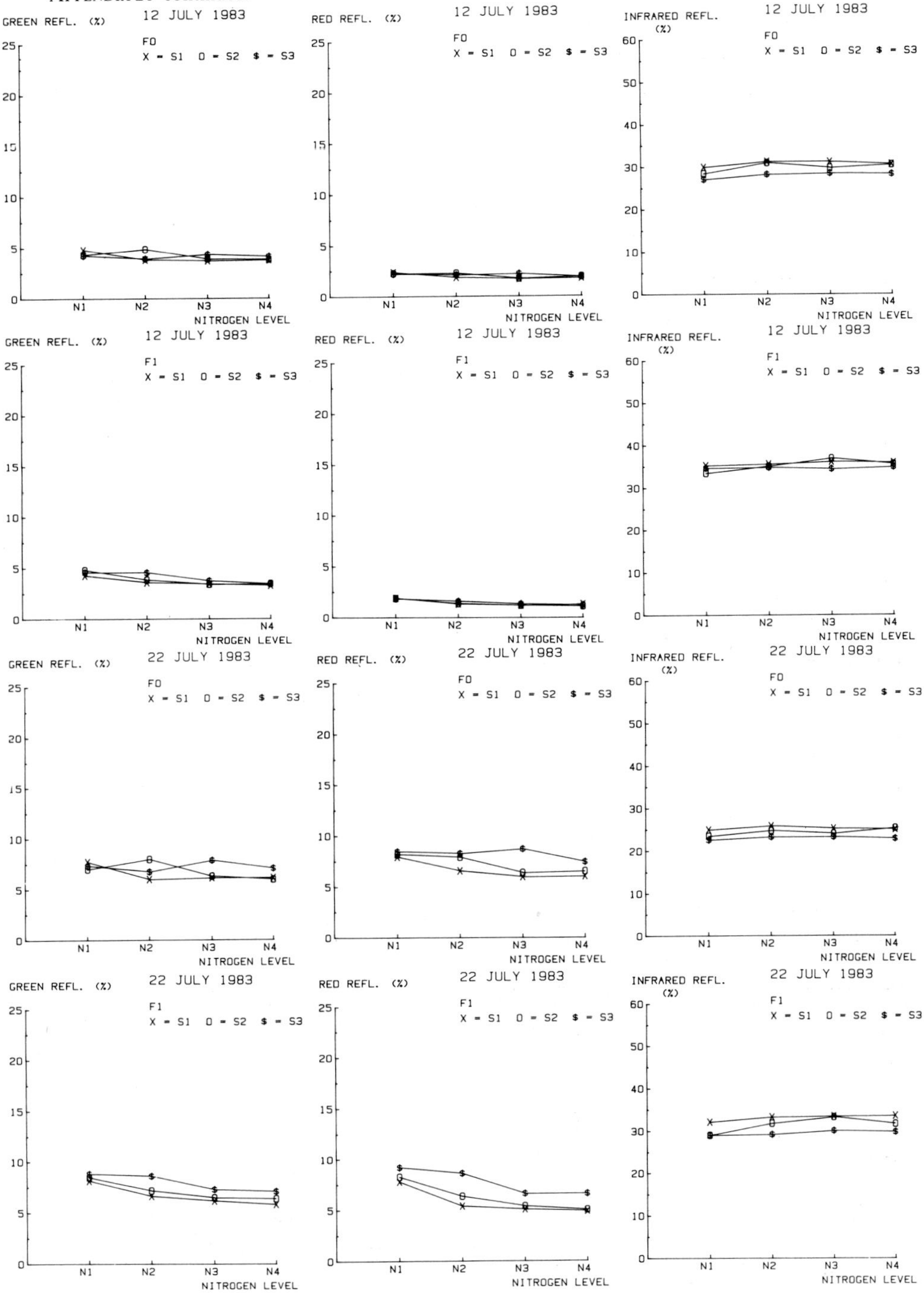

APPENDIX 21 continued.

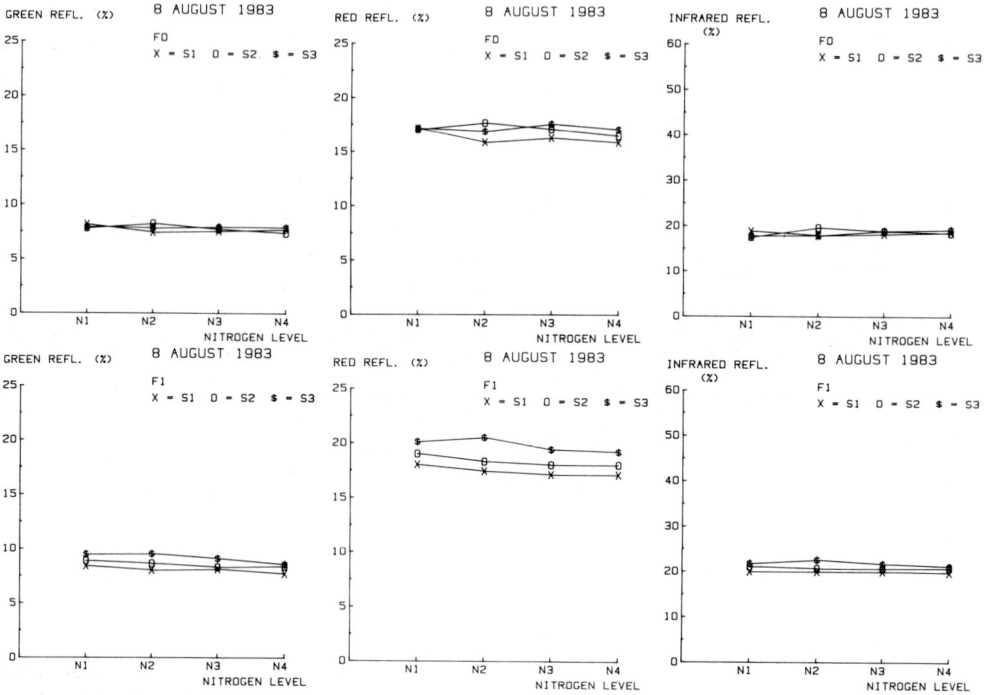

APPENDIX 22 Green and red and infrared reflectances per flying date. Field trial 95 in 1983.

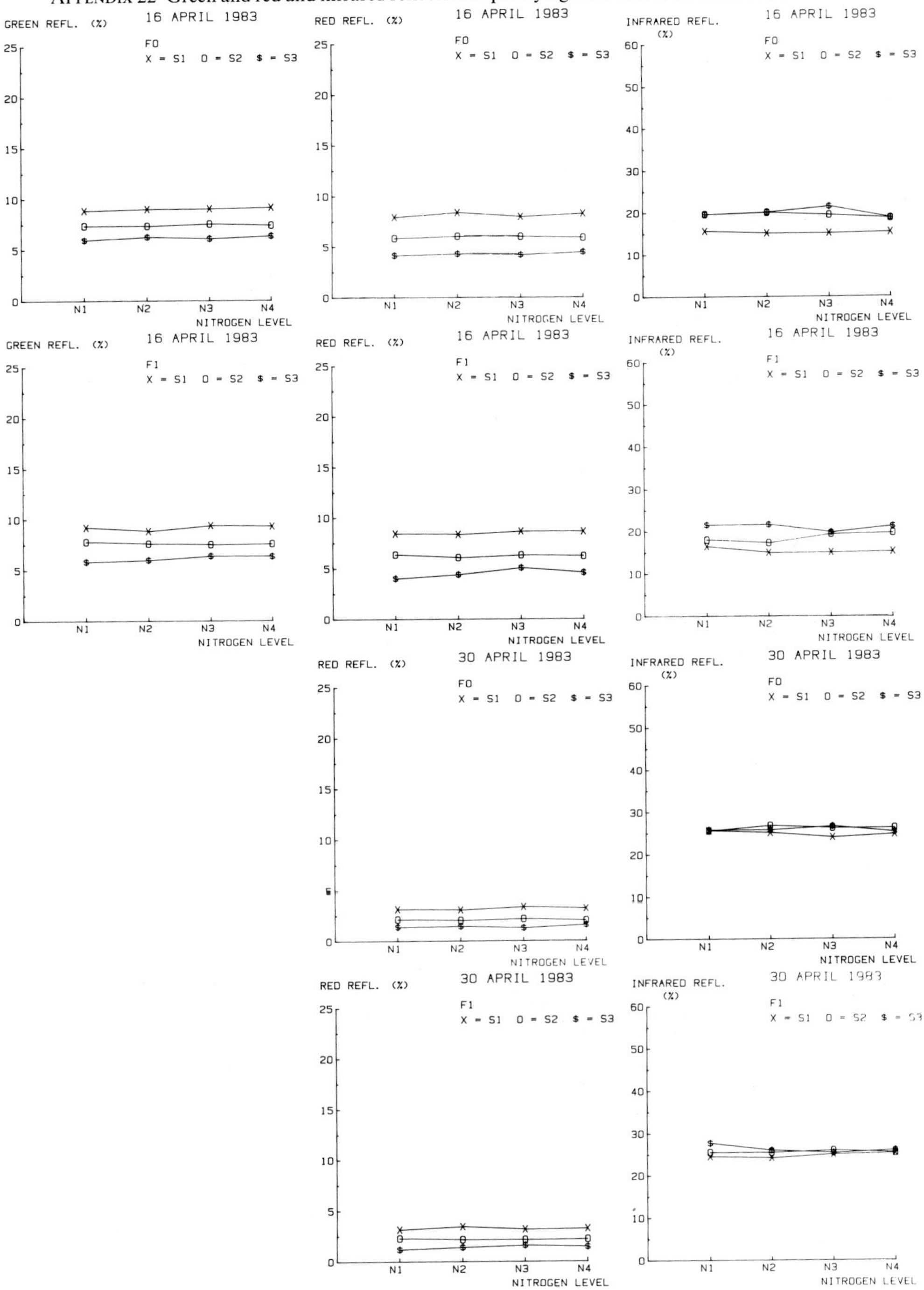

APPENDIX 22 continued.

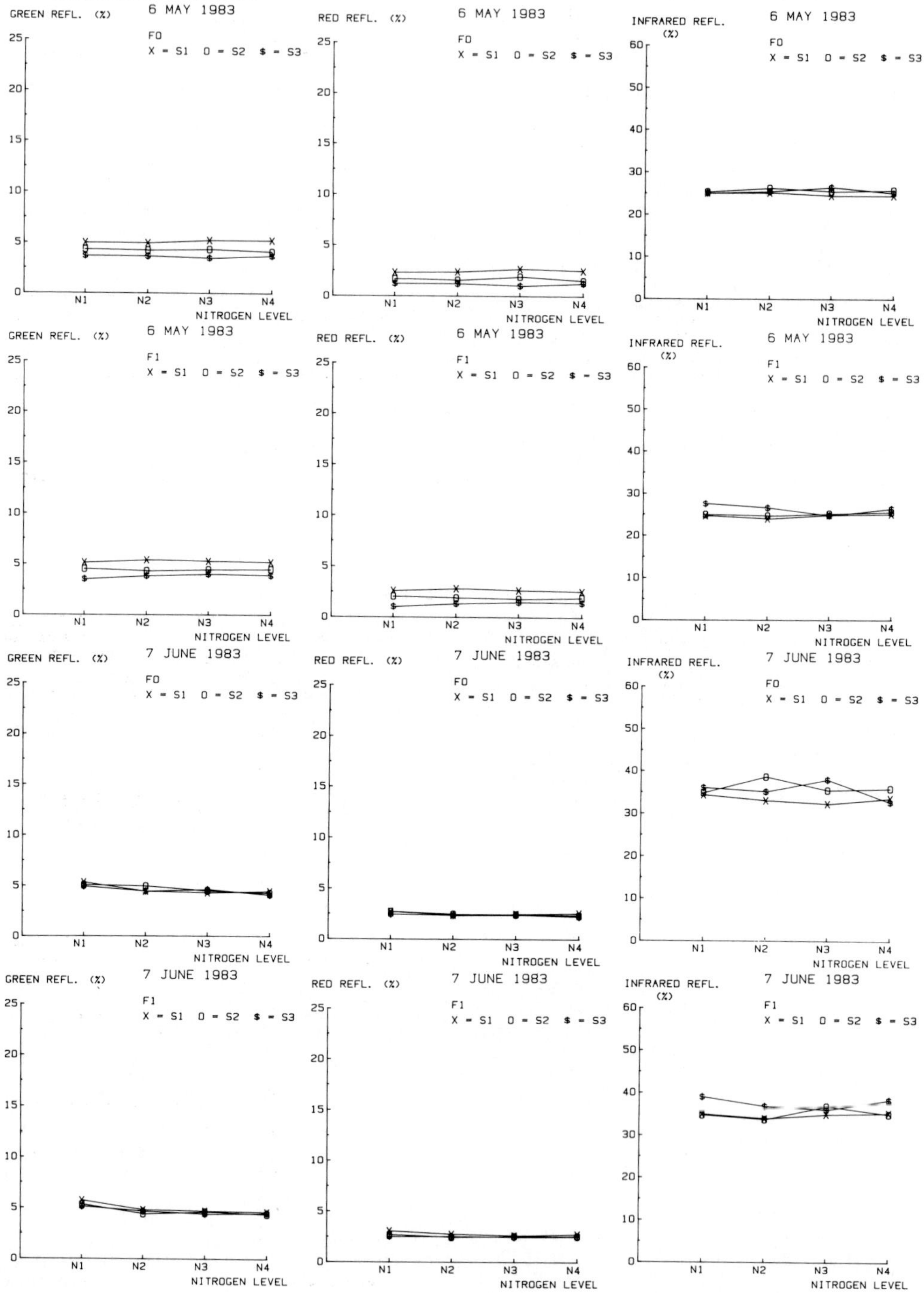

APPENDIX 22 continued.

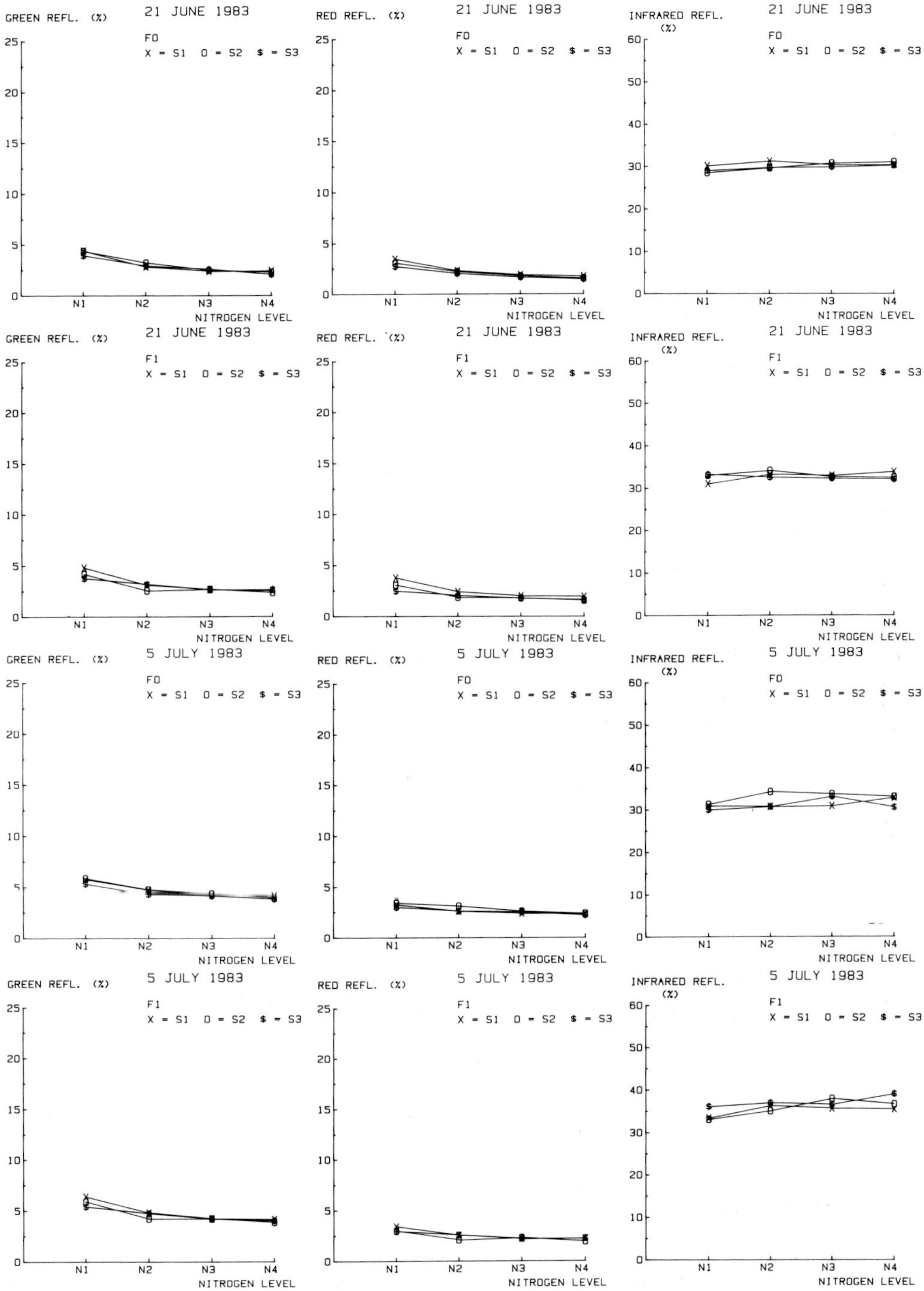

APPENDIX 22 continued.

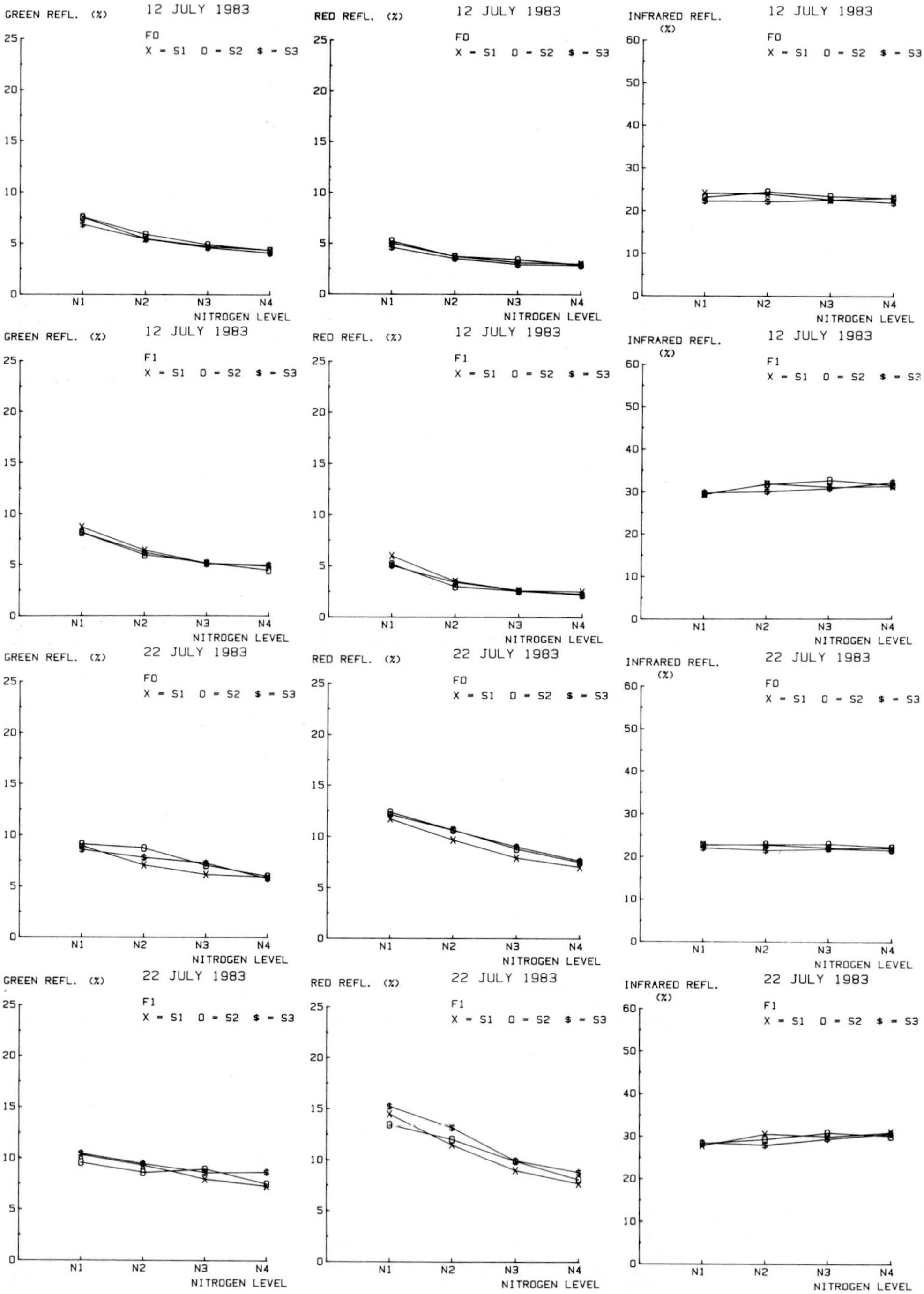

APPENDIX 22 continued.

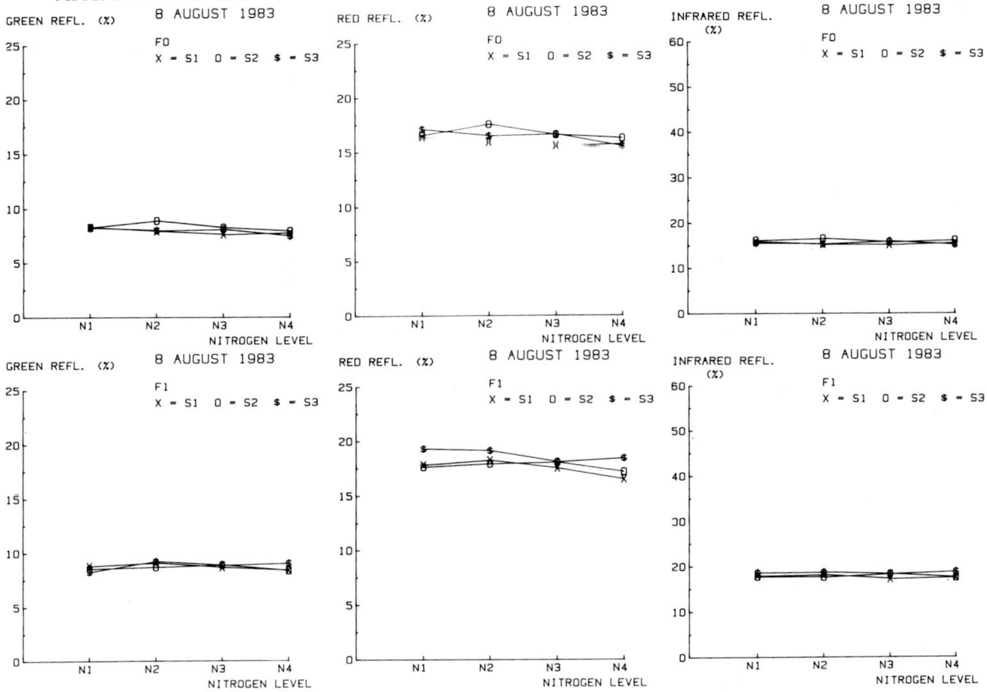

APPENDIX 23: Green and red reflectances (%): means, CVs and critical levels in testing for treatment effects, obtained on consecutive missions. Field trial 116 in 1983 (40 plots).
Green Reflectance

| date | mean | CV | critical level in testing: | | |
|---|---|---|---|---|---|
| | | | interaction | sowing dates | nitrogen nutrition |
| 16 April | 20.1 | 0.018 | 0.057 | 0.000 | 0.035 |
| 30 April | – | – | – | – | – |
| 6 May | 13.2 | 0.032 | 0.000 | 0.001 | 0.000 |
| 30 May | 5.70 | 0.031 | 0.652 | 0.000 | 0.098 |
| 7 June | 6.83 | 0.054 | 0.046 | 0.001 | 0.001 |
| 21 June | 4.26 | 0.084 | 0.000 | 0.003 | 0.000 |
| 5 July | 4.05 | 0.103 | 0.034 | 0.001 | 0.780 |
| 12 July | 6.65 | 0.059 | 0.002 | 0.000 | 0.060 |
| 22 July | 12.5 | 0.064 | 0.210 | 0.000 | 0.014 |
| 8 August | 10.4 | 0.048 | 0.000 | 0.294 | 0.000 |

Red Reflectance

| date | mean | CV | critical level in testing: | | |
|---|---|---|---|---|---|
| | | | interaction | sowing dates | nitrogen nutrition |
| 16 April | 23.6 | 0.016 | 0.016 | 0.000 | 0.007 |
| 30 April | 10.7 | 0.037 | 0.004 | 0.001 | 0.134 |
| 6 May | 14.0 | 0.038 | 0.000 | 0.000 | 0.000 |
| 30 May | 4.30 | 0.069 | 0.885 | 0.001 | 0.004 |
| 7 June | 4.36 | 0.103 | 0.433 | 0.000 | 0.002 |
| 21 June | 1.98 | 0.136 | 0.001 | 0.007 | 0.000 |
| 5 July | 1.82 | 0.143 | 0.172 | 0.001 | 0.001 |
| 12 July | 4.56 | 0.101 | 0.050 | 0.000 | 0.006 |
| 22 July | 13.4 | 0.099 | 0.011 | 0.000 | 0.000 |
| 8 August | 17.3 | 0.033 | 0.000 | 0.014 | 0.000 |

APPENDIX 24

A. Green reflectance (%): means, CVs and critical levels in testing for treatment effects, obtained on consecutive missions. Field trial 92 in 1982.

| date | mean | CV | critical level in testing: | | | | | | |
|---|---|---|---|---|---|---|---|---|---|
| | | | ABC | AB | AC | BC | A | B | C |
| 28 April | 19.8 | 0.036 | 0.323 | 0.720 | 0.793 | 0.433 | 0.168 | 0.000 | 0.665 |
| 17 May | 9.34 | 0.037 | 0.763 | 0.089 | 0.046 | 0.439 | 0.314 | 0.000 | 0.694 |
| 27 May | 5.15 | 0.041 | 0.482 | 0.948 | 0.016 | 0.091 | 0.349 | 0.000 | 0.034 |
| 10 June | 2.94 | 0.070 | 0.305 | 0.043 | 0.598 | 0.631 | 0.633 | 0.619 | 0.952 |
| 17 June | 3.15 | 0.055 | 0.224 | 0.416 | 0.581 | 0.740 | 0.888 | 0.987 | 0.927 |
| 1 July | 3.07 | 0.043 | 0.196 | 0.744 | 0.840 | 0.950 | 0.601 | 0.825 | 0.008 |
| 13 July | 3.59 | 0.056 | 0.295 | 0.853 | 0.770 | 0.722 | 0.970 | 0.000 | 0.004 |
| 29 July | 5.05 | 0.049 | 0.637 | 0.213 | 0.352 | 0.609 | 0.033 | 0.000 | 0.000 |
| 11 August | 7.76 | 0.041 | 0.105 | 0.049 | 0.714 | 0.641 | 0.200 | 0.000 | 0.000 |

B. Red reflectance (%): means, CVs and critical levels in testing for treatment effects, obtained on consecutive missions. Field trial 92 in 1982.

| date | mean | CV | critical level in testing: | | | | | | |
|---|---|---|---|---|---|---|---|---|---|
| | | | ABC | AB | AC | BC | A | B | C |
| 28 April | 10.9 | 0.023 | 0.081 | 0.180 | 0.606 | 0.850 | 0.572 | 0.000 | 0.579 |
| 17 May | 7.74 | 0.052 | 0.454 | 0.041 | 0.181 | 0.986 | 0.124 | 0.000 | 0.153 |
| 27 May | 5.50 | 0.084 | 0.400 | 0.909 | 0.207 | 0.576 | 0.169 | 0.000 | 0.033 |
| 10 June | 2.22 | 0.040 | 0.066 | 0.853 | 0.723 | 0.497 | 0.138 | 0.494 | 0.040 |
| 17 June | 1.15 | 0.099 | 0.592 | 0.980 | 0.109 | 0.687 | 0.216 | 0.069 | 0.396 |
| 1 July | 1.07 | 0.095 | 0.068 | 0.256 | 0.010 | 0.079 | 0.580 | 0.025 | 0.024 |
| 13 July | 2.59 | 0.061 | 0.270 | 0.438 | 0.038 | 0.109 | 0.374 | 0.393 | 0.000 |
| 29 July | 5.13 | 0.075 | 0.153 | 0.661 | 0.187 | 0.186 | 0.327 | 0.000 | 0.000 |
| 11 August | 10.7 | 0.031 | 0.196 | 0.011 | 0.112 | 0.631 | 0.051 | 0.000 | 0.000 |

C. Infrared reflectance (%): means, CVs and critical levels in testing for treatment effects, obtained on consecutive missions. Field trial 92 in 1982.

| date | mean | CV | critical level in testing: | | | | | | |
|---|---|---|---|---|---|---|---|---|---|
| | | | ABC | AB | AC | BC | A | B | C |
| 28 April | 17.1 | 0.022 | 0.838 | 0.591 | 0.768 | 0.340 | 0.132 | 0.000 | 0.413 |
| 17 May | 17.0 | 0.029 | 0.838 | 0.055 | 0.424 | 0.616 | 0.331 | 0.000 | 0.244 |
| 27 May | 35.1 | 0.036 | 0.821 | 0.006 | 0.951 | 0.298 | 0.958 | 0.000 | 0.805 |
| 10 June | 36.8 | 0.035 | 0.068 | 0.091 | 0.281 | 0.342 | 0.027 | 0.000 | 0.293 |
| 17 June | 42.2 | 0.024 | 0.883 | 0.575 | 0.290 | 0.037 | 0.916 | 0.000 | 0.030 |
| 1 July | 46.7 | 0.036 | 0.885 | 0.149 | 0.246 | 0.105 | 0.317 | 0.000 | 0.939 |
| 13 July | 40.9 | 0.036 | 0.526 | 0.247 | 0.167 | 0.094 | 0.019 | 0.677 | 0.278 |
| 29 July | 21.3 | 0.036 | 0.042 | 0.708 | 0.314 | 0.534 | 0.003 | 0.234 | 0.000 |
| 11 August | 13.0 | 0.046 | 0.646 | 0.004 | 0.358 | 0.758 | 0.005 | 0.000 | 0.000 |

Appendix 25

A. Green reflectance (%): means, CVs and critical levels in testing for treatment effects, obtained on consecutive missions. Field trial 92 in 1983.

| date | mean | CV | critical level in testing: | | | | | | |
|---|---|---|---|---|---|---|---|---|---|
| | | | ABC | AB | AC | BC | A | B | C |
| 16 April | 7.62 | 0.039 | 0.964 | 0.547 | 0.865 | 0.046 | 0.442 | 0.000 | 0.295 |
| 30 April | – | – | – | – | – | – | – | – | – |
| 6 May | 5.26 | 0.117 | 0.438 | 0.745 | 0.691 | 0.480 | 0.408 | 0.000 | 0.958 |
| 7 June | 2.45 | 0.114 | 0.015 | 0.345 | 0.304 | 0.398 | 0.122 | 0.111 | 0.892 |
| 21 June | 1.95 | 0.143 | 0.011 | 0.333 | 0.629 | 0.627 | 0.160 | 0.030 | 0.001 |
| 5 July | 2.75 | 0.071 | 0.023 | 0.380 | 0.353 | 0.356 | 0.017 | 0.521 | 0.000 |
| 12 July | 3.97 | 0.098 | 0.016 | 0.336 | 0.083 | 0.195 | 0.026 | 0.041 | 0.000 |
| 22 July | 7.03 | 0.096 | 0.052 | 0.396 | 0.045 | 0.073 | 0.054 | 0.000 | 0.000 |
| 8 August | 8.17 | 0.050 | 0.263 | 0.002 | 0.495 | 0.538 | 0.022 | 0.000 | 0.004 |

B. Red reflectance (%): means, CVs and critical levels in testing for treatment effects, obtained on consecutive missions. Field trial 92 in 1983.

| date | mean | CV | critical level in testing: | | | | | | |
|---|---|---|---|---|---|---|---|---|---|
| | | | ABC | AB | AC | BC | A | B | C |
| 16 April | 6.11 | 0.068 | 0.744 | 0.691 | 0.486 | 0.145 | 0.882 | 0.000 | 0.476 |
| 30 April | 2.79 | 0.139 | 0.626 | 0.357 | 0.738 | 0.262 | 0.437 | 0.000 | 0.821 |
| 6 May | 2.01 | 0.226 | 0.590 | 0.802 | 0.628 | 0.676 | 0.537 | 0.000 | 0.965 |
| 7 June | 1.47 | 0.085 | 0.182 | 0.481 | 0.756 | 0.045 | 0.837 | 0.000 | 0.909 |
| 21 June | 1.69 | 0.067 | 0.011 | 0.123 | 0.626 | 0.020 | 0.250 | 0.019 | 0.000 |
| 5 July | 1.83 | 0.132 | 0.105 | 0.803 | 0.773 | 0.585 | 0.006 | 0.083 | 0.000 |
| 12 July | 1.72 | 0.101 | 0.182 | 0.245 | 0.117 | 0.024 | 0.003 | 0.049 | 0.000 |
| 22 July | 6.96 | 0.093 | 0.152 | 0.373 | 0.005 | 0.104 | 0.012 | 0.000 | 0.000 |
| 8 August | 17.6 | 0.042 | 0.370 | 0.001 | 0.485 | 0.611 | 0.019 | 0.000 | 0.012 |

C. Infrared reflectance (%): means, CVs and critical levels in testing for treatment effects, obtained on consecutive missions. Field trial 92 in 1983.

| date | mean | CV | critical level in testing: | | | | | | |
|---|---|---|---|---|---|---|---|---|---|
| | | | ABC | AB | AC | BC | A | B | C |
| 16 April | 20.6 | 0.059 | 0.008 | 0.109 | 0.831 | 0.349 | 0.319 | 0.000 | 0.056 |
| 30 April | 26.2 | 0.033 | 0.638 | 0.591 | 0.345 | 0.877 | 0.691 | 0.683 | 0.839 |
| 6 May | 23.3 | 0.036 | 0.238 | 0.594 | 0.618 | 0.683 | 0.399 | 0.000 | 0.868 |
| 7 June | 26.1 | 0.072 | 0.223 | 0.005 | 0.131 | 0.408 | 0.932 | 0.000 | 0.703 |
| 21 June | 32.0 | 0.076 | 0.152 | 0.623 | 0.550 | 0.991 | 0.012 | 0.148 | 0.792 |
| 5 July | 37.7 | 0.042 | 0.037 | 0.003 | 0.383 | 0.851 | 0.005 | 0.319 | 0.057 |
| 12 July | 32.3 | 0.032 | 0.257 | 0.009 | 0.364 | 0.301 | 0.009 | 0.000 | 0.001 |
| 22 July | 27.8 | 0.036 | 0.152 | 0.034 | 0.019 | 0.095 | 0.009 | 0.000 | 0.000 |
| 8 August | 19.6 | 0.050 | 0.126 | 0.007 | 0.395 | 0.747 | 0.021 | 0.004 | 0.843 |

APPENDIX 26

A. Green reflectance (%): means, CVs and critical levels in testing for treatment effects, obtained on consecutive missions. Field trial 95 in 1983.

| date | mean | CV | critical level in testing: | | | | | | |
|---|---|---|---|---|---|---|---|---|---|
| | | | ABC | AB | AC | BC | A | B | C |
| 16 April | 7.58 | 0.040 | 0.515 | 0.334 | 0.479 | 0.362 | 0.417 | 0.000 | 0.130 |
| 30 April | – | – | – | – | – | – | – | – | – |
| 6 May | 4.39 | 0.075 | 0.464 | 0.972 | 0.816 | 0.899 | 0.354 | 0.000 | 0.846 |
| 7 June | 4.68 | 0.073 | 0.354 | 0.249 | 0.666 | 0.568 | 0.214 | 0.148 | 0.000 |
| 21 June | 3.02 | 0.131 | 0.419 | 0.192 | 0.696 | 0.173 | 0.184 | 0.304 | 0.000 |
| 5 July | 4.62 | 0.069 | 0.437 | 0.091 | 0.562 | 0.181 | 0.261 | 0.005 | 0.000 |
| 12 July | 5.77 | 0.079 | 0.940 | 0.046 | 0.189 | 0.671 | 0.066 | 0.206 | 0.000 |
| 22 July | 8.13 | 0.095 | 0.225 | 0.075 | 0.502 | 0.546 | 0.010 | 0.125 | 0.000 |
| 8 August | 8.41 | 0.049 | 0.272 | 0.002 | 0.838 | 0.636 | 0.020 | 0.187 | 0.003 |

B. Red reflectance (%): means, CVs and critical levels in testing for treatment effects, obtained on consecutive missions. Field trial 95 in 1983.

| date | mean | CV | critical level in testing: | | | | | | |
|---|---|---|---|---|---|---|---|---|---|
| | | | ABC | AB | AC | BC | A | B | C |
| 16 April | 6.25 | 0.054 | 0.402 | 0.728 | 0.080 | 0.415 | 0.182 | 0.000 | 0.162 |
| 30 April | 2.24 | 0.126 | 0.462 | 0.718 | 0.832 | 0.855 | 0.653 | 0.000 | 0.860 |
| 6 May | 1.85 | 0.162 | 0.245 | 0.796 | 0.759 | 0.765 | 0.212 | 0.000 | 0.737 |
| 7 June | 2.51 | 0.056 | 0.288 | 0.070 | 0.645 | 0.148 | 0.096 | 0.000 | 0.000 |
| 21 June | 2.15 | 0.123 | 0.752 | 0.220 | 0.675 | 0.019 | 0.233 | 0.000 | 0.000 |
| 5 July | 2.56 | 0.134 | 0.670 | 0.031 | 0.697 | 0.737 | 0.017 | 0.682 | 0.000 |
| 12 July | 3.51 | 0.119 | 0.883 | 0.075 | 0.002 | 0.514 | 0.083 | 0.070 | 0.000 |
| 22 July | 10.3 | 0.072 | 0.638 | 0.126 | 0.012 | 0.647 | 0.009 | 0.000 | 0.000 |
| 8 August | 17.2 | 0.038 | 0.111 | 0.003 | 0.944 | 0.479 | 0.011 | 0.000 | 0.000 |

C. Infrared reflectance (%): means, CVs and critical levels in testing for treatment effects, obtained on consecutive missions. Field trial 95 in 1983.

| date | mean | CV | critical level in testing: | | | | | | |
|---|---|---|---|---|---|---|---|---|---|
| | | | ABC | AB | AC | BC | A | B | C |
| 16 April | 18.3 | 0.083 | 0.224 | 0.116 | 0.278 | 0.844 | 0.493 | 0.000 | 0.926 |
| 30 April | 25.6 | 0.041 | 0.156 | 0.264 | 0.550 | 0.547 | 0.601 | 0.000 | 0.867 |
| 6 May | 25.3 | 0.044 | 0.092 | 0.071 | 0.225 | 0.879 | 0.981 | 0.002 | 0.927 |
| 7 June | 35.4 | 0.086 | 0.410 | 0.172 | 0.593 | 0.847 | 0.119 | 0.028 | 0.955 |
| 21 June | 31.4 | 0.053 | 0.387 | 0.712 | 0.758 | 0.923 | 0.003 | 0.514 | 0.383 |
| 5 July | 33.8 | 0.060 | 0.221 | 0.022 | 0.697 | 0.829 | 0.000 | 0.163 | 0.011 |
| 12 July | 27.0 | 0.037 | 0.491 | 0.263 | 0.001 | 0.123 | 0.001 | 0.007 | 0.028 |
| 22 July | 25.9 | 0.037 | 0.673 | 0.685 | 0.001 | 0.151 | 0.000 | 0.014 | 0.033 |
| 8 August | 16.8 | 0.048 | 0.462 | 0.014 | 0.986 | 0.909 | 0.001 | 0.099 | 0.837 |

APPENDIX 27: Results of LAI and reflectance measurements in field trial 100 in 1983.
A. LAI

| date | treatment | | | |
|---|---|---|---|---|
| | Z1N1 | Z1N4 | Z2N1 | Z2N4 |
| 10 May | 1.18 | 1.53 | 0.00 | 0.00 |
| 31 May | 1.68 | 2.84 | 0.60 | 0.85 |
| 9 June | 2.82 | 3.61 | 1.37 | 1.70 |
| 21 June | 2.40 | 3.33 | 2.06 | 2.87 |
| 5 July | 1.79 | 3.11 | 2.15 | 2.98 |
| 12 July | – | – | 1.70 | 2.38 |
| 22 July | – | – | 1.36 | 1.63 |
| 8 August | 0.00 | 0.00 | 0.00 | 0.00 |

B. Green reflectance (%)

| date | treatment | | | |
|---|---|---|---|---|
| | Z1N1 | Z1N4 | Z2N1 | Z2N4 |
| 6 May | 12.8 | 10.5 | 14.8 | 15.2 |
| 30 May | 6.17 | 5.93 | 6.98 | 6.79 |
| 7 June | 6.02 | 4.48 | 9.30 | 8.69 |
| 21 June | 4.62 | 2.52 | 6.42 | 5.60 |
| 5 July | 4.64 | 4.91 | 3.81 | 2.66 |
| 12 July | 7.63 | 8.71 | 4.83 | 4.24 |
| 22 July | 13.9 | 15.3 | 9.48 | 9.01 |
| 8 August | 9.06 | 10.7 | 10.7 | 9.14 |

C. Red reflectance (%)

| date | treatment | | | |
|---|---|---|---|---|
| | Z1N1 | Z1N4 | Z2N1 | Z2N4 |
| 6 May | 12.8 | 9.48 | 17.8 | 17.7 |
| 30 May | 5.11 | 4.23 | 7.19 | 6.75 |
| 7 June | 3.69 | 2.28 | 7.30 | 5.89 |
| 21 June | 2.47 | 0.89 | 3.36 | 2.49 |
| 5 July | 2.75 | 2.00 | 1.89 | 0.85 |
| 12 July | 6.42 | 5.97 | 3.82 | 2.43 |
| 22 July | 18.6 | 19.3 | 7.35 | 6.00 |
| 8 August | 15.3 | 17.9 | 16.9 | 14.7 |

APPENDIX 27 continued.
D. Infrared reflectance (%)

| date | treatment | | | |
|---|---|---|---|---|
| | Z1N1 | Z1N4 | Z2N1 | Z2N4 |
| 6 May | 25.8 | 27.5 | 14.0 | 14.3 |
| 30 May | 33.6 | 41.0 | 20.7 | 23.7 |
| 7 June | 40.2 | 47.4 | 35.4 | 39.6 |
| 21 June | 35.0 | 42.6 | 39.1 | 42.5 |
| 5 July | 33.5 | 44.4 | 33.7 | 39.2 |
| 12 July | 30.8 | 41.1 | 30.3 | 38.4 |
| 22 July | 33.2 | 40.3 | 37.5 | 44.5 |
| 8 August | 20.3 | 25.5 | 21.7 | 22.1 |